Matrix methods for engineers and scientists

Inverse $\quad AA^{-1} = I \quad$ exists is $\quad \det A \neq 0$

$A^{-1} = \frac{1}{\det A}$ adj A

solution of $Ax=b \Rightarrow x = A^{-1}b$

$\begin{bmatrix} 1 & 0 \\ 0 & 1 \end{bmatrix} A^{-1}$ PG. 66 $\quad$ Gauss - Jordan elimination to find <u>inverse</u>

Summarization on pg 68

PG 77 $\quad$ Rank $\quad \begin{array}{l} R(A) = R(A,b) = n \\ R(A) = R(A,b) < n \\ R(A) < R(A,b) \end{array} \quad \begin{array}{l} 1 \text{ solution} \\ \infty \\ 0 \end{array}$

P. 84 $\quad$ Normal form

P. 90 $\quad$ General solution $\quad x = x_n + x_p$

P 97 $\quad$ Linear dependence

Characteristic Polynomial : Pg 109 - 111

Hermitian & Symmetric matrices : Pg 112

Cayley - Hamilton Theorem : Pg 115

Similiarity : Pg 122 $\qquad \rightarrow B = P^{-1}AP$

Diagonalization : Pg 124

DET $A = $ product of Eigenvalues

TR $A = $ sum $\quad " \quad "$

IN $P^{-1}AP$: eigenvalues are along diagonal

Matrix methods for engineers and scientists

Stephen Barnett
Reader in Engineering Mathematics
School of Mathematics
University of Bradford

McGRAW-HILL Book Company (UK) Limited

London · New York · St. Louis · San Francisco · Auckland
Bogotá · Düsseldorf · Johannesburg · Lisbon · Lucerne
Madrid · Mexico · Montreal · New Delhi · Panama · Paris
San Juan · São Paulo · Singapore · Sydney · Tokyo · Toronto

Published by
McGraw-Hill Book Company (UK) Limited
Maidenhead · Berkshire · England

British Library Cataloguing in Publication Data
Barnett, Stephen
  Matrix methods for engineers and scientists.
  1. Matrices
  I.Title
  512.9'43   QA188   78-40555

  ISBN 0-07-084084-9

1 2 3 4 5 A W 8 0 7 9 8

Printed and bound in Great Britain by A. Wheaton & Co., Ltd., Exeter

Contents

v

Preface

Matrix algebra is an essential and practical tool for virtually all users of mathematics. Moreover, students usually find the work a welcome relief from the intricacies of calculus. This book presents an up-to-date, straightforward course for students of engineering and science; it is based on many years' experience both of teaching matrix methods, and of active research in applied matrix problems. My treatment of the material is non-abstract and relatively informal, with the emphasis on topics of value for applications. I have aimed at clarity of explanations, so that results are often justified using simple cases, rather than by general proofs. After an introductory chapter showing how the concept of a matrix arises in applications, the next three chapters provide a basic course on algebra and properties of matrices, unique solution of linear equations, determinant, and inverse. The second half of the book comprises a further course, covering rank and non-unique solution of equations, eigenvalues and eigenvectors, quadratic forms, and matrix functions. The material is mainly standard, but I have included some interesting and useful topics (e.g., companion matrix, Kronecker product) which are often omitted from elementary books, and illustrative applications are described throughout. From the computational aspect, a recurrent theme is the simple but powerful method of gaussian elimination, which in various guises is used for the solution of equations (Chapters 3 and 5), triangular decomposition (Chapter 3), evaluation of determinant and inverse (Chapter 4) and of rank (Chapter 5), and for testing the sign nature of a quadratic form (Chapter 7). In addition, some iterative numerical schemes are described in Chapter 6.

The problems in the text are essential for understanding the subject matter, and should be attempted as they occur (it is a truism that mathematics cannot be learnt by reading about it, any more than one can learn to play a musical instrument by watching a performer). The exercises at the end of each chapter are equally important but tend to be somewhat harder than the problems, so not all need be tried at a first reading. I have also used the exercises to introduce further results and applications.

Finally, I must acknowledge that besides its role in applications, matrix theory also possesses its own particular fascination which has given me much pleasure over the years. I have been greatly stimulated by books such as Mirsky (1963) and the first edition of Bellman (1970), and I therefore hope that in a more modest way this present volume may succeed in conveying some of the appeal of the subject.

I am grateful to those of my colleagues who made helpful suggestions during the preparation of the manuscript (Tim Cronin, John Grant, and John Maroulas deserve special mention), and to Wynne Smith for her excellent and painstaking production of the typescript. I also wish to thank Professor J. B. Helliwell for making available resources of the University of Bradford, Steve Teale for his work on the drawings, and John Maroulas for help in correcting the proofs.

Stephen Barnett
Bradford, June 1978

Notation

$A = [a_{ij}]$	matrix having element a_{ij} in row i, column j
$a = [a_1, a_2, \ldots, a_n]$	row n-vector having a_i as ith component
A^T	transpose of A, $= [a_{ji}]$
$b = [b_1, b_2, \ldots, b_n]^T$	column n-vector
$\bar{A}$	complex conjugate of A, $= [\bar{a}_{ij}]$
A^*	conjugate transpose of A, $= (\bar{A})^T$
$\text{diag}\,[d_{11}, d_{22}, \ldots, d_{nn}]$	$n \times n$ diagonal matrix, $d_{ij} = 0$, $i \neq j$
I_n	$n \times n$ unit matrix, $= \text{diag}\,[1, 1, \ldots, 1]$
$A \otimes B$	Kronecker product of A and B
$\text{tr}\,(A)$	trace of $n \times n$ matrix A, $= a_{11} + a_{22} + \cdots + a_{nn}$
$\dot{x}$	$dx/dt = [dx_1/dt, \ldots, dx_n/dt]$
$\ddot{x}$	d^2x/dt^2
$(Ri), (Cj), (Lk)$	row i, column j, line k (of A)
$\det A, |A|$	determinant of square matrix A
PDi	property of determinants, number i (Section 4.1.2)
$\text{adj}\,A$	adjoint of A
A^{-1}	inverse of A
$R(A)$	rank of A
$\|x\|$	euclidean norm of vector x, $= (\Sigma |x_i|^2)^{1/2}$
$\|A\|$	euclidean norm of A, $= (\Sigma\Sigma |a_{ij}|^2)^{1/2}$
(4.39)	refers to equation number (4.39) in Chapter 4

1. How matrices arise

In this chapter we describe a few of the very many areas of applications in which matrices are used, the choice of topics being deliberately diverse.

Example 1.1 Suppose that the prices (in some monetary units) of four kinds of canned foods F_1, F_2, F_3, F_4 at three different supermarkets S_1, S_2, S_3 are as given in Table 1.1.

	F_1	F_2	F_3	F_4
S_1	17	7	11	21
S_2	15	9	13	19
S_3	18	8	15	19

Table 1.1

Thus for example the price of item F_3 in supermarket S_2 is 13. The total cost of buying one can of each food at S_1 is $17+7+11+21 = 56$, and similarly 56 at S_2, 60 at S_3.

The rectangular array of numbers in Table 1.1 is called a *matrix*, in this case having three *rows* and four *columns*. An array like this will arise whenever there are two sets (above, the foods and the supermarkets) whose members are linked by a set of numbers (above, the prices).

Example 1.2 Table 1.2 gives distances (in miles) between four American cities, and is an example of a familiar feature of many road maps.

	Chicago	New York City	San Francisco	Washington DC
Chicago	0	841	2212	704
New York City	841	0	3033	224
San Francisco	2212	3033	0	2835
Washington DC	704	224	2835	0

Table 1.2

1

The array here has the same number of rows as columns, and is called *square*. The line of zeros forms the *principal diagonal* of this square array. Notice also that the numbers in Table 1.2 are symmetric with respect to this diagonal, e.g., the distance from New York to San Francisco is given either by the number in row 2, column 3, or by the number in row 3, column 2. We shall see that such *symmetric* matrices have interesting properties.

Example 1.3 Figure 1.1, called a *network*, represents connections between two airports A_1, A_2 in one country with airports B_1, B_2, B_3 in a second country. The number on each linking line gives the number of different airlines flying on that route, e.g., there are two airlines offering

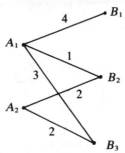

Fig. 1.1   Airport connections for Example 1.3.

flights from A_2 to B_2. In tabular form the information can be presented as follows:

$$\begin{array}{c} \quad B_1 \ B_2 \ B_3 \\ \begin{array}{c} A_1 \\ A_2 \end{array} \begin{bmatrix} 4 & 1 & 3 \\ 0 & 2 & 2 \end{bmatrix} \end{array} \qquad (1.1)$$

This time we have enclosed the array within square brackets, and this is the standard notation for matrices.

Example 1.4 It can be shown that the currents i_1, i_2, i_3 in the electric circuit represented in Fig. 1.2 satisfy the equations

$$(R_1 + R_4 + R_5)i_1 - R_4i_2 - R_5i_3 = E_1$$
$$- R_4i_1 + (R_2 + R_4 + R_6)i_2 - R_6i_3 = E_2 \qquad (1.2)$$
$$- R_5i_1 - R_6i_2 + (R_3 + R_5 + R_6)i_3 = E_3$$

These equations are to be solved for the three unknowns i_1, i_2, i_3, the values of the other parameters being known. The Eqs (1.2) are called *linear* because no powers or products of the i's are involved. The

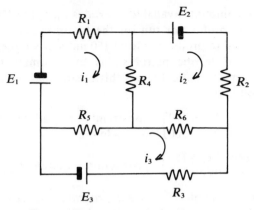

Fig. 1.2 Electric circuit for Example 1.4.

information in (1.2) can be presented in a more concise way using matrix notation, as follows:

$$\begin{bmatrix} (R_1+R_4+R_5) & -R_4 & -R_5 \\ -R_4 & (R_2+R_4+R_6) & -R_6 \\ -R_5 & -R_6 & (R_3+R_5+R_6) \end{bmatrix} \begin{bmatrix} i_1 \\ i_2 \\ i_3 \end{bmatrix} = \begin{bmatrix} E_1 \\ E_2 \\ E_3 \end{bmatrix} \quad (1.3)$$

At this stage we regard (1.3) merely as an alternative way of *writing* (1.2), having some advantage in economy of symbols since the coefficients in the equations have been recorded separately from the unknowns. However we shall see in Chapter 2 that a fundamental significance can be attached to this representation. Sets of linear equations arise in many applications and their solution forms a major theme of this book.

Example 1.5 An oil refinery makes two grades of petrol, 'supreme' and 'terrific', from two crude oils c_1 and c_2. Two possible blending processes can be used, and the inputs and outputs for a single production run are shown in Table 1.3. Thus if x_1, x_2 are the numbers of production runs of processes 1 and 2 respectively, then the amount of crude c_1 used is $3x_1 + 5x_2$ and the amount of supreme petrol produced is $4x_1 + 5x_2$, and so on.

	Input		Output	
	crude c_1	crude c_2	supreme	terrific
Process 1	3	4	4	6
Process 2	5	2	5	3

Table 1.3

3

The maximum amounts available of crudes c_1, c_2 are 170 and 200 units respectively. The profits per run for the two processes are 2.1 and 2.3 units. It is required to produce at least 150 units of supreme petrol and at least 110 units of terrific petrol so as to maximize the total profit $2.1x_1 + 2.3x_2$. In mathematical terms this means that we must determine $x_1 \geq 0$, $x_2 \geq 0$ satisfying

$$\left. \begin{array}{l} 3x_1 + 5x_2 \leq 170 \\ 4x_1 + 2x_2 \leq 200 \end{array} \right\} \quad \begin{array}{l} \text{constraints on availability} \\ \text{of crude oils} \end{array}$$

$$\left. \begin{array}{l} 4x_1 + 5x_2 \geq 150 \\ 6x_1 + 3x_2 \geq 110 \end{array} \right\} \quad \text{production requirements} \qquad (1.4)$$

This is a simple example of a *linear programming* (LP) problem. The linear inequalities can easily be converted into equations. For example, if we define the amount of unused crude c_1 as x_3, then

$$x_3 = 170 - 3x_1 - 5x_2$$

and the first inequality in (1.4) implies $x_3 \geq 0$. Therefore this inequality can be replaced by the equation

$$3x_1 + 5x_2 + x_3 = 170 \qquad (1.5)$$

in which, like x_1 and x_2, the new variable x_3 must be non-negative.

Similarly, the other three constraints for the problem become

$$4x_1 + 2x_2 + x_4 = 200$$
$$4x_1 + 5x_2 - x_5 = 150 \qquad (1.6)$$
$$6x_1 + 3x_2 - x_6 = 110$$

with x_4, x_5, x_6 all to be non-negative. In (1.5) and (1.6) there are now more unknowns (six) than equations (four). This is a typical feature of LP problems, which constitute an important field of application of matrices.

In matrix form (1.5) and (1.6) can be written

$$\begin{bmatrix} 3 & 5 & 1 & 0 & 0 & 0 \\ 4 & 2 & 0 & 1 & 0 & 0 \\ 4 & 5 & 0 & 0 & -1 & 0 \\ 6 & 3 & 0 & 0 & 0 & -1 \end{bmatrix} \begin{bmatrix} x_1 \\ x_2 \\ x_3 \\ x_4 \\ x_5 \\ x_6 \end{bmatrix} = \begin{bmatrix} 170 \\ 200 \\ 150 \\ 110 \end{bmatrix}$$

✳ **Example 1.6**✳ Let P be a point in the plane having cartesian coordinates x_1 and x_2, and let O be the origin, as shown in Fig. 1.3a. Suppose that OP is rotated in an anticlockwise direction through an angle θ so that P moves to $P' \equiv (x_1', x_2')$ as in Fig. 1.3b. Since $OP = OP' (= r$, say) we have

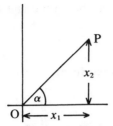

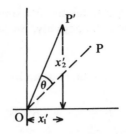

Fig. 1.3a   Point P in Example 1.6.   Fig. 1.3b   Point P′ in Example 1.6.

$x_1 = r \cos \alpha$, $x_2 = r \sin \alpha$ and

$$x_1' = r \cos (\theta + \alpha)$$
$$= r \cos \theta \cos \alpha - r \sin \theta \sin \alpha$$
$$= (\cos \theta)x_1 - (\sin \theta)x_2$$

and similarly

$$x_2' = (\sin \theta)x_1 + (\cos \theta)x_2$$

or in matrix terms

$$\begin{bmatrix} x_1' \\ x_2' \end{bmatrix} = \begin{bmatrix} \cos \theta & -\sin \theta \\ \sin \theta & \cos \theta \end{bmatrix} \begin{bmatrix} x_1 \\ x_2 \end{bmatrix} \qquad (1.7)$$

The change of coordinates from x_1, x_2 to x_1', x_2' is an example of a *transformation* (or *mapping*). These are useful in geometry.

Example 1.7   Consider a mechanical system composed of two masses lying on a smooth table, connected to each other and to a fixed support by springs and dampers as shown in Fig. 1.4. The displacements of the masses from equilibrium are x_1 and x_2, and their velocities are x_3 and x_4 respectively. Assuming that the springs obey Hooke's law, and the dampers exert forces proportional to velocity, Newton's equations of motion can be written (ignoring u):

$$\left. \begin{array}{ll} \text{For } m_1: & m_1\dot{x}_3 = -k_1(x_1 - x_2) - d_1(x_3 - x_4) \\ \text{For } m_2: & m_2\dot{x}_4 = k_1(x_1 - x_2) + d_1(x_3 - x_4) - k_2x_2 - d_2x_4 \end{array} \right\} \qquad (1.8)$$

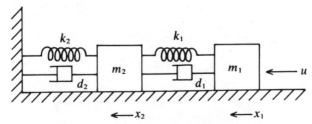

Fig. 1.4   Mass–spring system for Example 1.7.

5

where $(\cdot)$ denotes $d(\ )/dt$, and k_1, k_2, d_1, d_2 are the spring and damping coefficients respectively. Since

$$\dot{x}_1 = x_3, \dot{x}_2 = x_4 \tag{1.9}$$

we can write Eqs (1.8) and (1.9) in the combined form

$$\begin{bmatrix} \dot{x}_1 \\ \dot{x}_2 \\ \dot{x}_3 \\ \dot{x}_4 \end{bmatrix} = \begin{bmatrix} 0 & 0 & 1 & 0 \\ 0 & 0 & 0 & 1 \\ -k_1/m_1 & k_1/m_1 & -d_1/m_1 & d_1/m_1 \\ k_1/m_2 & -(k_1+k_2)/m_2 & d_1/m_2 & -(d_1+d_2)/m_2 \end{bmatrix} \begin{bmatrix} x_1 \\ x_2 \\ x_3 \\ x_4 \end{bmatrix} \tag{1.10}$$

constituting a set of linear *differential* equations.

If the right-hand mass were given an initial push then the system would perform unforced oscillations. Alternatively, it may be required to determine how a force u should be applied to m_1 so as to make the masses move in a desired fashion – this is a problem of *control theory*. In both areas matrices play a very important role, some aspects of which will be dealt with in later chapters. Some books on applications are listed in the Bibliography.

Exercises

1.1 The network in Fig. 1.5 represents roads connecting three cities in country D to three cities in country E, and then to two cities in country F. The connections between the cities in countries D and E can be described by the following matrix:

$$\begin{array}{c} \\ d_1 \\ d_2 \\ d_3 \end{array} \begin{array}{c} e_1 \ \ e_2 \ \ e_3 \\ \begin{bmatrix} 1 & 1 & 0 \\ 1 & 0 & 1 \\ 1 & 1 & 0 \end{bmatrix} \end{array}$$

where the 1's indicate pairs of cities which are connected, and the zeros pairs which are not. Write down the matrix for road connections between countries E and F.

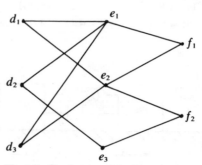

Fig. 1.5 Road network for Exercise 1.1.

6

1.2 Suppose that for the network in Fig. 1.1, onward flights from country B to country C are as shown in Fig. 1.6.

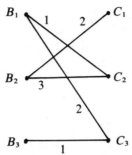

Fig. 1.6 Onward airport connections for Exercise 1.2.

(a) Write down the matrix giving information on flights between B and C.
(b) By combining Figs 1.1 and 1.6, obtain the matrix giving the numbers of different ways of flying from airports in A to those in C.

1.3 Two people, A, B, play a game in which they each toss a coin simultaneously. The rules are as follows:
if both coins come up heads A pays B five pence; if both come up tails B pays A seven pence; if A's shows a head and B's a tail then B pays A two pence; and if A's shows a tail and B's a head then A pays B three pence.

Represent this information in matrix form. How would the rules have to be modified so as to make this matrix symmetric?

1.4 The minimum daily requirements of an adult person are 2000 units of proteins, 2500 units of fats, and 8500 units of carbohydrates. The contents of certain foods are given in Table 1.4 in units per gram.

	Proteins	Fats	Carbohydrates
Food 1	2	1	16
Food 2	8	2	0
Food 3	0	28	0
Food 4	2	9	1

Table 1.4

Let x_i, $i = 1, 2, 3, 4$, represent the weights bought of each food. Determine the inequalities that represent the condition that the minimum daily requirements be met or exceeded. Convert these inequalities into linear equations, and hence express the problem in matrix terms.

If in addition the daily requirements are to be met at minimum cost, then this is an LP problem.

7

1.5 Using the notation of Example 1.6, what do the following transformations represent in geometrical terms?

(a) $\begin{bmatrix} x_1' \\ x_2' \end{bmatrix} = \begin{bmatrix} 1 & 0 \\ 0 & -1 \end{bmatrix} \begin{bmatrix} x_1 \\ x_2 \end{bmatrix}$
(b) $\begin{bmatrix} x_1' \\ x_2' \end{bmatrix} = \begin{bmatrix} 2 & 0 \\ 0 & 2 \end{bmatrix} \begin{bmatrix} x_1 \\ x_2 \end{bmatrix}$

(c) $\begin{bmatrix} x_1' \\ x_2' \\ x_3' \end{bmatrix} = \begin{bmatrix} \cos\theta & -\sin\theta & 0 \\ \sin\theta & \cos\theta & 0 \\ 0 & 0 & 1 \end{bmatrix} \begin{bmatrix} x_1 \\ x_2 \\ x_3 \end{bmatrix}$

2. Basic algebra of matrices

2.1 Definitions

The first step in the development of matrix algebra is to denote an array by a single letter – almost invariably an upper case ('capital') letter is used, for example

$$A = \begin{bmatrix} 1 & 7 & 5 \\ 3 & -1 & 2 \end{bmatrix} \tag{2.1}$$

In some books bold type, A, is used for matrices. The numbers in the array are called the *elements* of A. The standard notation for elements is to use the same letter but lower case, and with two suffices which describe the position of an element. For example, in Eq. (1.1) $a_{11} = 1$, $a_{12} = 7$, $a_{21} = 3$, etc. Generally a_{ij} denotes the element in *row i* and *column j* of the array, and is called the *i,j* element: thus

$$A = \begin{bmatrix} a_{11} & a_{12} & a_{13} & \cdot & \cdot & \cdot & a_{1n} \\ a_{21} & a_{22} & a_{23} & \cdot & \cdot & \cdot & a_{2n} \\ \cdot & \cdot & \cdot & \cdot & \cdot & \cdot & \cdot \\ \cdot & \cdot & \cdot & \cdot & a_{ij} & \cdot & \cdot \\ \cdot & \cdot & \cdot & \cdot & \cdot & \cdot & \cdot \\ a_{m1} & a_{m2} & a_{m3} & \cdot & \cdot & \cdot & a_{mn} \end{bmatrix} \leftarrow \text{row } i \tag{2.2}$$

$$\uparrow$$
$$\text{column } j$$

The matrix A in (2.2) has m rows and n columns, and is said to have *dimensions* $m \times n$, or simply to be an $m \times n$ matrix. A useful shorthand for (2.2) is

$$A = [a_{ij}], \quad i = 1, 2, \ldots, m; \quad j = 1, 2, \ldots, n \tag{2.3}$$

When $m = n$ the matrix is called *square* of *order n*; otherwise A is called *rectangular*. In this book the elements will be real or complex numbers. When all the a_{ij} are real then A is called a *real* matrix; if some or all of the a_{ij} are complex then A is a *complex* matrix. For a square matrix, the diagonal from the top left corner to the bottom right corner is called the *principal* diagonal.

Two matrices A and B are *equal* if they have the same dimensions and all their elements in corresponding positions are identical, i.e., $a_{ij} = b_{ij}$, for all possible i and j.

9

One of the crucial advantages of matrix notation is that very often properties and theorems can be obtained by direct manipulation of matrices denoted by single letters, without involving use of the actual elements in the arrays.

Problem 2.1 Consider the matrix

$$B = \begin{bmatrix} 1 & 3 \\ 7 & -1 \\ 5 & 4 \\ 2 & -6 \end{bmatrix} \tag{2.4}$$

What are: (a) the dimensions of B; (b) the elements b_{12}, b_{31}, b_{13}?

Problem 2.2 Write down the matrix A in (2.3) when

(a) $m = 2$, $n = 3$, $a_{ij} = 2i - j$; (b) $m = 3$, $n = 3$, $a_{ij} = |i - j|$.

2.2 Basic operations

2.2.1 Addition

Two matrices A and B can be added together only if they have the same dimensions, and then the elements in corresponding positions are added, i.e., if $A + B = C$ then

$$c_{ij} = a_{ij} + b_{ij} \tag{2.5}$$

Clearly the order of addition does not matter, i.e., $A + B = B + A$, so addition is *commutative*.

2.2.2 Multiplication by a scalar

If k is a constant then the product kA is formed by multiplying every element of A by k. In our shorthand notation this can be written

$$k[a_{ij}] = [ka_{ij}] \tag{2.6}$$

The two rules (2.5) and (2.6) can be combined together. For example, taking $k = -1$ in (2.6), the rule for subtraction is that the i,j element of $A - B = A + (-1)B$ is

$$a_{ij} + (-1)b_{ij} = a_{ij} - b_{ij} \tag{2.7}$$

that is, corresponding elements are subtracted.

Example 2.1 If A is the matrix in Eq. (2.1) and

$$B = \begin{bmatrix} 2 & 4 & 2 \\ 0 & 3 & -1 \end{bmatrix} \tag{2.8}$$

then

$$A + B = \begin{bmatrix} (1+2) & (7+4) & (5+2) \\ (3+0) & (-1+3) & (2-1) \end{bmatrix}$$

$$= \begin{bmatrix} 3 & 11 & 7 \\ 3 & 2 & 1 \end{bmatrix}$$

$$2A = \begin{bmatrix} 2 & 14 & 10 \\ 6 & -2 & 4 \end{bmatrix}$$

$$3B = \begin{bmatrix} 6 & 12 & 6 \\ 0 & 9 & -3 \end{bmatrix}$$

$$2A - 3B = \begin{bmatrix} (2-6) & (14-12) & (10-6) \\ (6-0) & (-2-9) & (4+3) \end{bmatrix}$$

$$= \begin{bmatrix} -4 & 2 & 4 \\ 6 & -11 & 7 \end{bmatrix}$$

After a little practice intermediate steps can be missed out.

Problem 2.3 If

$$A = \begin{bmatrix} 1 & -1 \\ 0 & 4 \\ 4 & 2 \\ 3 & -8 \end{bmatrix} \qquad (2.9)$$

and B is the matrix in Eq. (2.4), determine (a) $A + B$, (b) $A - B$, (c) $2B$, (d) $2B - 4A$.

It is obvious that for A in Eq. (2.1)

$$A - A = \begin{bmatrix} 0 & 0 & 0 \\ 0 & 0 & 0 \end{bmatrix} \qquad (2.10)$$

and this will clearly be true for any matrix. A matrix like that in (2.10) having all its elements zero is called, naturally enough, a *zero matrix* (sometimes, a *null* matrix).

Notice also that (2.6) can be applied in reverse so as to remove a factor common to all the elements of a matrix. For example,

$$\begin{bmatrix} 3 & 9 \\ -6 & 12 \end{bmatrix} = 3\begin{bmatrix} 1 & 3 \\ -2 & 4 \end{bmatrix}$$

The rules presented so far can be extended to more than two matrices in an obvious fashion. For example, if A, B, C are three matrices having the same dimensions then their sum is

$$A + B + C = A + (B + C)$$
$$= (A + B) + C, \text{etc.}$$

11

This states that the result of adding A, B, and C is independent of the order in which it is done. In formal terms, matrix addition is thus *associative*. Similarly, $2A + 3B + 5C = (2A + 3B) + 5C$, etc.

Problem 2.4 If A, B, C are three matrices having the same dimensions, and

$$A + C = B + C$$

show by considering elements on both sides of the equation that $A = B$.

Example 2.2 Consider an equation

$$X + A = B \tag{2.11}$$

which is to be solved for the matrix X, given A and B. Add $-A$ on to both sides of Eq. (2.11) to give

$$X + A + (-A) = B + (-A)$$
$$X + (A - A) = B - A$$
$$X + 0 = B - A$$

where 0 denotes the zero matrix, so

$$X = B - A \tag{2.12}$$

since $X + 0 = X$. Thus, although A, B, X stand for arrays of numbers, we have been able to manipulate them as single entities. In this case the solution (2.12) of Eq. (2.11) is the same as would be obtained if A, B, X were scalars. In the next section we begin to break away from ordinary algebra.

2.2.3 Multiplication of two matrices

Example 2.3 Return to the supermarket problem described in Example 1.1. Let x_i be the number of cans purchased of food F_i, $i = 1, 2, 3, 4$. Then from Table 1.1 the total cost at supermarket S_1 would be $17x_1 + 7x_2 + 11x_3 + 21x_4$, and similarly for S_2 and S_3. We can write these products in the following form, as in Example 1.4:

$$\underbrace{\begin{bmatrix} 17 & 7 & 11 & 21 \\ 15 & 9 & 13 & 19 \\ 18 & 8 & 15 & 19 \end{bmatrix}}_{A} \underbrace{\begin{bmatrix} x_1 \\ x_2 \\ x_3 \\ x_4 \end{bmatrix}}_{X} = \begin{bmatrix} 17x_1 + 7x_2 + 11x_3 + 21x_4 \\ 15x_1 + 9x_2 + 13x_3 + 19x_4 \\ 18x_1 + 8x_2 + 15x_3 + 19x_4 \end{bmatrix} \tag{2.13}$$

The elements of the matrix on the right-hand side of (2.13) give the costs of the purchases at S_1, S_2, and S_3 respectively. It therefore seems natural to define the *product* of A and X as in (2.13): each of the rows of A is multiplied term-by-term with the elements of X. For this to work it is clear that the number of *columns* of A must be the same as the number of *rows* in X.

Problem 2.5 Three types of food f_1, f_2, f_3 have vitamin content in units per kilogram given in Table 2.1.

	f_1	f_2	f_3
Vit. A	3	2	4
Vit. B	5	7	9

Table 2.1

Express the vitamin content of 5 kg of f_1, 3 kg of f_2, and 7 kg of f_3 as a matrix product, and evaluate it. If the costs per kilogram of the three foods are 75, 90 and 80 pence respectively, express the total cost as a matrix product, and evaluate it.

If the second matrix in the product has more than one column we simply multiply each of its columns in turn by A, using the same rule. For example, if

$$\underset{A}{\begin{bmatrix} a_{11} & a_{12} \\ a_{21} & a_{22} \end{bmatrix}} \underset{B}{\begin{bmatrix} b_{11} & b_{12} \\ b_{21} & b_{22} \end{bmatrix}} = \underset{C}{\begin{bmatrix} (a_{11}b_{11} + a_{12}b_{21}) & (a_{11}b_{12} + a_{12}b_{22}) \\ (a_{21}b_{11} + a_{22}b_{21}) & (a_{21}b_{12} + a_{22}b_{22}) \end{bmatrix}} \quad (2.14)$$

then we write $AB = C$. The *first* column of C in (2.14) is obtained by multiplying the *first* column of B by A, as in (2.13); the *second* column of C is obtained by multiplying the *second* column of B by A.

Example 2.4 If

$$A = \begin{bmatrix} 1 & 2 \\ 3 & 4 \end{bmatrix}, \quad B = \begin{bmatrix} -1 & 4 \\ 2 & -3 \end{bmatrix} \quad (2.15)$$

then

$$AB = \begin{bmatrix} \{1 \times (-1) + 2 \times 2\} & \{1 \times 4 + 2 \times (-3)\} \\ \{3 \times (-1) + 4 \times 2\} & \{3 \times 4 + 4 \times (-3)\} \end{bmatrix}$$

$$= \begin{bmatrix} 3 & -2 \\ 5 & 0 \end{bmatrix} \quad (2.16)$$

The general rule is as follows:

If $A = [a_{ij}]$ is $m \times n$ and $B = [b_{ij}]$ is $n \times p$ then $C = AB$ is $m \times p$ and

13

c_{ij} = term-by-term product of the ith row of A with the jth column of B

$$= [a_{i1}a_{i2} \cdots a_{in}] \begin{bmatrix} b_{1j} \\ b_{2j} \\ \vdots \\ b_{nj} \end{bmatrix} \tag{2.17}$$

$$= a_{i1}b_{1j} + a_{i2}b_{2j} + \cdots + a_{in}b_{nj} \tag{2.18}$$

$$= \sum_{k=1}^{n} a_{ik}b_{kj} \tag{2.19}$$

Thus AB is constructed by multiplying the first column of B by each of the rows of A in turn, giving the first column of AB; this is repeated with the second column of B, and so on. Notice again that the product AB is defined only if the number of *columns* of A is equal to the number of *rows* of B, and A and B are then said to be *conformable* for multiplication. The dimensions of the resulting product can be found by the simple rule:

$$\underset{(m \times n)}{A} \quad \underset{(n \times p)}{B} = \underset{(m \times p)}{C} \tag{2.20}$$

Example 2.4 (continued)   For A and B in Eq. (2.15) we have

$$BA = \begin{bmatrix} \{(-1) \times 1 + 4 \times 3\} & \{(-1) \times 2 + 4 \times 4\} \\ \{2 \times 1 + (-3) \times 3\} & \{2 \times 2 + (-3) \times 4\} \end{bmatrix}$$

$$= \begin{bmatrix} 11 & 14 \\ -7 & -8 \end{bmatrix}$$

which has no elements in common with AB in (2.16). This illustrates the general fact that $AB \neq BA$, i.e., matrix multiplication is *not* commutative – a fundamental departure from ordinary algebra.

In AB the matrix B is said to be multiplied *on the left*, or *premultiplied* by A; in BA the matrix B is multiplied *on the right*, or *postmultiplied* by A. In general we shall have to specify which is meant – the expression 'the product of A and B' is too vague.

Example 2.5   If A is the matrix in Eq. (2.1) and

$$B = \begin{bmatrix} -1 & 6 & 7 \\ 4 & 5 & 3 \\ 3 & 0 & 4 \end{bmatrix}$$

14

then using the rule for multiplication

$$AB = \begin{bmatrix} \{1 \times (-1) + 7 \times 4 + 5 \times 3\} & (1 \times 6 + 7 \times 5 + 5 \times 0) \\ \{3 \times (-1) + (-1) \times 4 + 2 \times 3\} & \{3 \times 6 + (-1) \times 5 + 2 \times 0\} \end{bmatrix}$$
$$\begin{matrix} (1 \times 7 + 7 \times 3 + 5 \times 4) \\ \{3 \times 7 + (-1) \times 3 + 2 \times 4\} \end{matrix} \Big]$$

$$= \begin{bmatrix} 42 & 41 & 48 \\ -1 & 13 & 26 \end{bmatrix}$$

and the dimensions of AB can be obtained from (2.20): $(\underline{2} \times 3)(3 \times \underline{3})$. Notice that BA does not exist: consider $(3 \times \underline{3})(\underline{2} \times 3)$ – the two underlined dimensions are not equal.

Problem 2.6 Determine where possible $A + B$, AB, and BA for each of the following pairs:

(a) $A = \begin{bmatrix} 0 & 1 \\ 1 & 1 \end{bmatrix}$, $\qquad B = \begin{bmatrix} 0 & -1 \\ 1 & 0 \end{bmatrix}$

(b) $A = \begin{bmatrix} 1 & 2 & 3 \\ 1 & 3 & 6 \end{bmatrix}$, $\qquad B = \begin{bmatrix} 1 & 1 & 1 \\ 1 & 2 & 3 \end{bmatrix}$

(c) $A = \begin{bmatrix} 2 & -1 \\ 1 & 0 \\ -3 & 4 \end{bmatrix}$, $\qquad B = \begin{bmatrix} 1 & -2 & -5 \\ 3 & 4 & 0 \end{bmatrix}$

Problem 2.7 If A has m rows and $m + 5$ columns, B has n rows and $11 - n$ columns, and both AB and BA exist, what are the values of m and n?

Example 2.6 An important special matrix can now be introduced. Let

$$I = \begin{bmatrix} 1 & 0 \\ 0 & 1 \end{bmatrix} \tag{2.21}$$

and let A be the arbitrary 2×2 matrix in (2.14). Then with $B = I$ in (2.14) it is easy to verify that

$$AI = A \tag{2.22}$$

and similarly

$$IA = A \tag{2.23}$$

Since I behaves like the number 1 in ordinary algebra it is called the *unit matrix* (or *identity* matrix), being defined in general as the $n \times n$ matrix having 1's along its principal diagonal and zeros everywhere else. The results in Eqs (2.22) and (2.23) then hold for any $n \times n$ matrix A. We shall often write I_n to emphasize that the order is n.

15

Example 2.7 It can turn out that AB does equal BA, as is easily verified when

$$A = \begin{bmatrix} 1 & 2 \\ -2 & 1 \end{bmatrix}, \qquad B = \begin{bmatrix} 3 & 4 \\ -4 & 3 \end{bmatrix}$$

In this case A and B are said to *commute* with each other. In view of Eqs (2.22) and (2.23), I_n is a matrix which commutes with *all* $n \times n$ matrices.

Powers of a matrix are defined in an obvious fashion:

$$A^2 = AA, \qquad A^3 = AA^2, \qquad A^4 = AA^3, \ldots \tag{2.24}$$

It is important to realize that the dimension requirement in (2.20) implies that (2.24) is meaningful only if A is square. Since

$$A^3 = AAA = A^2 A$$

and so on, it follows that all powers of A commute with each other, and $A^k A^l = A^{k+l}$ for positive integers k and l.

Problem 2.8 If

$$A = \begin{bmatrix} 1 & -1 & 1 \\ 2 & -1 & 0 \\ 1 & 0 & 0 \end{bmatrix}$$

calculate A^2 and verify that $A^2 A = AA^2 = I_3$.

Problem 2.9 If

$$A = \begin{bmatrix} 3 & -1 & -1 \\ 1 & 1 & -1 \\ 1 & -1 & 1 \end{bmatrix} \tag{2.25}$$

calculate A^2 and verify that $A^2 - 3A + 2I_3 \equiv 0$.

Notice that it easily follows from the definition that matrix multiplication is *associative*, i.e.,

$$A(BC) = (AB)C \tag{2.26}$$

so we can simply write ABC without brackets. The dimensions of this product are shown below:

$$\begin{array}{cccc} A & B & C & = & D \\ (m \times n) & (n \times p) & (p \times q) & & (m \times q) \end{array}$$

16

It is also easy to show that multiplication is *distributive* with respect to addition, i.e.,

$$A(B + C) = AB + AC \tag{2.27}$$

The results (2.26) and (2.27) mean that we can deal with brackets in the same way as for ordinary algebra.

Problem 2.10 If A is $a_1 \times a_2$, B is $b_1 \times b_2, \ldots, Z$ is $z_1 \times z_2$, what are the conditions for the product $ABC \ldots YZ$ to exist, and what are its dimensions?

Problem 2.11 Test (2.26) with A, B as in Problem 2.6c and

$$C = \begin{bmatrix} 2 & 1 \\ 0 & -4 \\ 1 & 2 \end{bmatrix}$$

We now consider some further ways in which matrix multiplication differs from multiplication of scalars.

Problem 2.12 If

$$A = \begin{bmatrix} 2 & -3 & -5 \\ -1 & 4 & 5 \\ 1 & -3 & -4 \end{bmatrix}, \quad B = \begin{bmatrix} -1 & 3 & 5 \\ 1 & -3 & -5 \\ -1 & 3 & 5 \end{bmatrix}, \quad C = \begin{bmatrix} 2 & -2 & -4 \\ -1 & 3 & 4 \\ 1 & -2 & -3 \end{bmatrix}$$

show that (a) $AB = BA = 0$; (b) $AC = A$, $CA = C$, and hence show that $ACB = CBA$.

Part (a) of Problem 2.12 illustrates the fact that a product AB can equal the zero matrix even if neither A nor B is itself zero. This is in sharp contrast to ordinary algebra, where the result that a product can only be zero if one of the factors is zero is applied almost instinctively. For example, to solve a quadratic equation like

$$x^2 - 3x + 2 = 0 \tag{2.28}$$

we factorize as

$$(x - 1)(x - 2) = 0 \tag{2.29}$$

and then conclude that one of the factors in (2.29) must be zero, so the solution of (2.28) is $x = 1$ or $x = 2$. However, in attempting to solve a corresponding matrix equation

$$X^2 - 3X + 2I = 0 \tag{2.30}$$

17

where X is an $n \times n$ matrix, then although this factorizes into

$$(X - I)(X - 2I) = 0$$

(verify this!) we cannot assume that the only solutions of (2.30) are $X = I$ or $X = 2I$. Indeed, when X is a 3×3 matrix then a solution of (2.30) is the matrix in (2.25).

Another aspect of the preceding remarks is that if $AC = AD$, it cannot necessarily be inferred that $C = D$ (i.e., the A cannot be 'cancelled') for it is possible to have

$$AC - AD = A(C - D) = 0$$

even though $C - D$ may not equal zero. This is illustrated by part (b) of Problem 2.12, where $AC = A = AI_3$ but $C \neq I_3$.

Problem 2.13 If A and B are two $n \times n$ matrices, expand the product $(A - B)(A + B)$. Under what conditions is this equal to $A^2 - B^2$?

Problem 2.14 Expand $(A + B)^2$, $(A + B)^3$ for two arbitrary $n \times n$ matrices A and B. What is the condition for the usual binomial theorem expressions to be obtained?

Problem 2.15 Give an example of a 2×3 matrix A and a 3×2 matrix B for which $AB = I_2$, but $BA \neq I_3$.

We close this section with some more applications of matrix multiplication, emphasizing further that the definition is not a mathematical abstraction.

Example 2.8 Consider the network in Fig. 1.5, showing roads connecting three sets of cities. The matrices representing connections between countries D and E, and E and F are respectively

$$A = \begin{array}{c} \begin{array}{ccc} e_1 & e_2 & e_3 \end{array} \\ \begin{bmatrix} 1 & 1 & 0 \\ 1 & 0 & 1 \\ 1 & 1 & 0 \end{bmatrix} \begin{array}{c} d_1 \\ d_2, \\ d_3 \end{array} \end{array} \qquad B = \begin{array}{c} \begin{array}{cc} f_1 & f_2 \end{array} \\ \begin{bmatrix} 1 & 0 \\ 1 & 1 \\ 0 & 1 \end{bmatrix} \begin{array}{c} e_1 \\ e_2 \\ e_3 \end{array} \end{array} \qquad (2.31)$$

where $a_{ij} = 1$ if d_i is connected to e_j, and $a_{ij} = 0$ otherwise, and similarly for matrix B. Consulting Fig. 1.5, it is easily seen that the number of routes connecting d_1 and f_1 is 2 (via e_1 or via e_2). Proceeding in this way, the matrix giving numbers of routes between D and F is found to be

$$\begin{array}{c} \begin{array}{cc} f_1 & f_2 \end{array} \\ \begin{bmatrix} 2 & 1 \\ 1 & 1 \\ 2 & 1 \end{bmatrix} \begin{array}{c} d_1 \\ d_2 \\ d_3 \end{array} \end{array} \qquad (2.32)$$

18

The reader can verify that the product AB of the matrices in (2.31) gives precisely the matrix in (2.32).

Example 2.9 If $f(x, y)$ is a function of two independent variables and a transformation into $f(u, v)$ is made, then the theory of partial differentiation gives the derivatives of f with respect to the new variables in the form

$$\frac{\partial f}{\partial u} = \frac{\partial f}{\partial x}\frac{\partial x}{\partial u} + \frac{\partial f}{\partial y}\frac{\partial y}{\partial u}$$

$$\frac{\partial f}{\partial v} = \frac{\partial f}{\partial x}\frac{\partial x}{\partial v} + \frac{\partial f}{\partial y}\frac{\partial y}{\partial v}$$

In matrix terms this becomes

$$\begin{bmatrix} f_u \\ f_v \end{bmatrix} = \begin{bmatrix} x_u & y_u \\ x_v & y_v \end{bmatrix}\begin{bmatrix} f_x \\ f_y \end{bmatrix}$$

where x_u denotes $\partial x/\partial u$, etc.

Example 2.10 Let $z_1 = a_1 + ib_1$ be an arbitrary complex number with a_1 and b_1 real and $i^2 = -1$. Form the matrix

$$A^{(1)} = \begin{bmatrix} a_1 & b_1 \\ -b_1 & a_1 \end{bmatrix} \tag{2.33}$$

and denote the relationship by $A^{(1)} \sim z_1$. It turns out that properties of z_1 can be interpreted in terms of properties of $A^{(1)}$. To see this, let z_2 be a second complex number with $A^{(2)} \sim z_2$. Then

$$A^{(1)} + A^{(2)} = \begin{bmatrix} (a_1 + a_2) & (b_1 + b_2) \\ -(b_1 + b_2) & (a_1 + a_2) \end{bmatrix}$$

$$\sim (a_1 + a_2) + i(b_1 + b_2) \tag{2.34}$$

and the number in (2.34) is just $z_1 + z_2$, which means that addition of complex numbers can be done by adding matrices in the form of that in (2.33).

This also applies for multiplication:

$$A^{(1)} A^{(2)} = \begin{bmatrix} a_1 & b_1 \\ -b_1 & a_1 \end{bmatrix}\begin{bmatrix} a_2 & b_2 \\ -b_2 & a_2 \end{bmatrix}$$

$$= \begin{bmatrix} (a_1 a_2 - b_1 b_2) & (a_1 b_2 + b_1 a_2) \\ -(a_1 b_2 + b_1 a_2) & (a_1 a_2 - b_1 b_2) \end{bmatrix}$$

$$\sim (a_1 a_2 - b_1 b_2) + i(a_1 b_2 + b_1 a_2) \tag{2.35}$$

and the complex number in (2.35) is precisely $z_1 z_2$.

Further aspects of this representation of complex numbers via 2×2 matrices in the form of (2.33) will be developed in Exercise 2.9 and Problems 4.1 and 6.3.

Problem 2.16 A square $n \times n$ matrix $D = [d_{ij}]$ is called *diagonal* if all elements off the principal diagonal are zero, i.e., $d_{ij} = 0$, $i \neq j$. This is written

$$D = \text{diag}[d_{11}, d_{22}, \ldots, d_{nn}] \qquad (2.36)$$

Prove that

$$D^2 = \text{diag}[d_{11}^2, d_{22}^2, \ldots, d_{nn}^2]$$

and obtain the expression for D^k, where k is a positive integer. Show also that any two $n \times n$ diagonal matrices commute with each other.

2.3 Transpose

2.3.1 Definition and properties

If A is the general $m \times n$ matrix in Eq. (2.2) then the $n \times m$ matrix obtained from A by interchanging the rows and columns is called the *transpose* of A, written A^T (in some books A').

Example 2.11 If A is the matrix in (2.9) then

$$A^T = \begin{bmatrix} 1 & 0 & 4 & 3 \\ -1 & 4 & 2 & -8 \end{bmatrix} \qquad (2.37)$$

since the first row in (2.9) becomes the first column in (2.37), the second row becomes the second column, and so on. (Similarly, the first row of (2.37) is the first column of (2.9), etc.)

In general, the i,j element a_{ij} of A becomes the j,i element of A^T. For example, the 2,3 element of (2.37) is the 3,2 element of (2.9). In particular when A is square, the elements on the principal diagonal of A^T are a_{ii}, the same as those on the principal diagonal of A.

It is obvious that transposing twice in succession returns any matrix to itself, i.e.,

$$(A^T)^T = A \qquad (2.38)$$

Similarly it follows immediately from the definition that

$$(A + B)^T = A^T + B^T \qquad (2.39)$$

$$(kA)^T = kA^T, \text{ for any scalar } k. \qquad (2.40)$$

It is not quite so easy to prove that if A and B are conformable for multiplication then

$$(AB)^\mathsf{T} = B^\mathsf{T} A^\mathsf{T} \tag{2.41}$$

Note the reversal of order between the two sides of (2.41).

Problem 2.17 Determine $B^\mathsf{T} A^\mathsf{T}$ for the matrices in (2.15), and hence test the validity of (2.41).

To see how (2.41) is proved in general, suppose A and B are both 3×3 and let $C = AB$. The i,j element of C^T is c_{ji}, which from (2.18) with i and j interchanged is

$$c_{ji} = a_{j1}b_{1i} + a_{j2}b_{2i} + a_{j3}b_{3i} \tag{2.42}$$

The i,j element of $B^\mathsf{T} A^\mathsf{T}$ is the term-by-term product of the ith row of B^T with the jth column of A^T. However, the ith *row* of B^T is equal to the ith *column* of B and the jth *column* of A^T is equal to the jth *row* of A. Thus the i,j element of $B^\mathsf{T} A^\mathsf{T}$ is

$$b_{1i}a_{j1} + b_{2i}a_{j2} + b_{3i}a_{j3}$$

which is identical to (2.42), showing $C^\mathsf{T} = B^\mathsf{T} A^\mathsf{T}$ as required. The argument is easily extended to general dimensions.

If

$$x = \begin{bmatrix} x_1 \\ x_2 \\ \vdots \\ x_n \end{bmatrix}, \quad y = \begin{bmatrix} y_1 \\ y_2 \\ \vdots \\ y_n \end{bmatrix} \tag{2.43}$$

these $n \times 1$ matrices are called *column n-vectors*, and x^T, y^T are *row n-vectors*. The elements x_i, y_i are called *components* of the vectors. Again, bold type, $\mathbf{x}$, $\mathbf{y}$, is often used for vectors. The product

$$x^\mathsf{T} y = x_1 y_1 + x_2 y_2 + \cdots + x_n y_n \tag{2.44}$$

is the *scalar* (or *inner*) product of x and y.

Problem 2.18 Prove, using (2.41), that $(ABC)^\mathsf{T} = C^\mathsf{T} B^\mathsf{T} A^\mathsf{T}$.

Problem 2.19 Let e_i denote the ith row of I_n. For example, if $n = 3$ then $e_2 = [0, 1, 0]$. If A is an arbitrary $n \times n$ matrix, what do the products $e_i A$ and $A e_i^\mathsf{T}$ represent?

21

If some or all the a_{ij} are complex numbers, then the *complex conjugate* $\bar{A}$ of A is the matrix obtained by replacing each a_{ij} by its conjugate, $\bar{a}_{ij}$. The *conjugate transpose* of A is

$$A^* = (\bar{A})^{\mathrm{T}} \tag{2.45}$$

and the order of the operations in (2.45) doesn't matter, i.e.,

$$A^* = (\overline{A^{\mathrm{T}}})$$

The results of (2.38), (2.39), and (2.41) still hold for conjugate transpose with superscript T replaced by *. Similarly, (2.40) holds if k is real, but if k is complex we have

$$(kA)^* = \bar{k}A^* \tag{2.46}$$

Example 2.12

$$A = \begin{bmatrix} 2+3i & 1+i \\ 2-i & 4i \end{bmatrix}, \qquad A^* = \begin{bmatrix} 2-3i & 2+i \\ 1-i & -4i \end{bmatrix} \tag{2.47}$$

Notice that any complex matrix can be written as

$$A = A_1 + iA_2 \tag{2.48}$$

where the elements of A_1 and A_2 are purely real. For example, for (2.47)

$$A = \begin{bmatrix} 2 & 1 \\ 2 & 0 \end{bmatrix} + i \begin{bmatrix} 3 & 1 \\ -1 & 4 \end{bmatrix}$$

2.3.2 Symmetric and hermitian matrices

We encountered the idea of a *symmetric* square matrix in Example 1.2. This can now be defined by

$$A^{\mathrm{T}} = A \tag{2.49}$$

which implies that $a_{ij} = a_{ji}$, for all i and j. Similarly A is *skew symmetric* if

$$A^{\mathrm{T}} = -A \tag{2.50}$$

which implies $a_{ij} = -a_{ji}$, for all i and j, so in particular all the diagonal elements a_{ii} are zero.

Example 2.13 The matrices

$$A = \begin{bmatrix} 1 & 3 & 7 \\ 3 & 4 & 2 \\ 7 & 2 & 0 \end{bmatrix}, \qquad B = \begin{bmatrix} 0 & 3 & 7 \\ -3 & 0 & 2 \\ -7 & -2 & 0 \end{bmatrix} \tag{2.51}$$

are respectively symmetric and skew symmetric.

An interesting fact is that any square matrix A can be expressed *uniquely* as

$$A = M + S \qquad (2.52)$$

where M is symmetric and S is skew symmetric. To show this, transpose both sides of (2.52):

$$A^T = (M + S)^T$$
$$= M^T + S^T$$
$$= M - S \qquad (2.53)$$

Adding (2.52) and (2.53) gives the *symmetric part* of A

$$M = \tfrac{1}{2}(A + A^T) \qquad (2.54)$$

and similarly by subtraction the *skew symmetric part* is

$$S = \tfrac{1}{2}(A - A^T) \qquad (2.55)$$

Problem 2.20 Determine M and S for the matrix A in (2.25).

Problem 2.21 Verify that the matrices of (2.54) and (2.55) are indeed symmetric and skew symmetric.

Problem 2.22 If A is any symmetric $n \times n$ matrix and P is an arbitrary $m \times n$ matrix, prove that PAP^T is symmetric.

Problem 2.23 Prove that the maximum number of different elements in an $n \times n$ symmetric matrix is $\tfrac{1}{2}n(n + 1)$. What is the maximum number for a skew symmetric matrix (ignoring signs and zeros)?

When A has complex elements, two further important definitions are used: If

$$A^* = A \qquad (2.56)$$

then A is called *hermitian* (after the French mathematician Hermite), and (2.56) implies $a_{ij} = \bar{a}_{ji}$; and if

$$A^* = -A \qquad (2.57)$$

then A is called *skew hermitian*, with $a_{ij} = -\bar{a}_{ji}$.

Example 2.14 The matrices

$$\begin{bmatrix} 2 & 1+i & 5-i \\ 1-i & 7 & i \\ 5+i & -i & -1 \end{bmatrix}, \quad \begin{bmatrix} 2i & 1+i & 5-i \\ -1+i & 7i & i \\ -5-i & i & -i \end{bmatrix}$$

are respectively hermitian and skew hermitian.

23

This example illustrates the general fact that for a hermitian matrix $a_{ii} = \bar{a}_{ii}$, so all a_{ii} are purely real; and similarly, all a_{ii} are purely imaginary for a skew hermitian matrix.

Important applications of the matrices introduced in this section will be studied in Chapter 7.

Problem 2.24   If A is any rectangular matrix prove that A^*A and AA^* are both hermitian.

Problem 2.25   Consider the case when A in (2.48) is hermitian. (a) Show that A_1 is symmetric and A_2 is skew symmetric. (b) Determine the condition to be satisfied by A_1 and A_2 for A^*A to be a real matrix.

Problem 2.26   If A is any skew hermitian matrix prove that iA and $-iA$ are both hermitian.

Problem 2.27   Obtain the generalization of Eqs (2.52), (2.54), and (2.55) when A is complex.

2.4   Partitioning and submatrices

It is often convenient to subdivide or *partition* a matrix into smaller blocks of elements.

Example 2.15   The following 3×5 matrix is partitioned into four blocks

$$A = \left[\begin{array}{ccc|cc} 1 & 0 & 2 & 3 & 5 \\ 2 & 1 & 4 & 3 & 0 \\ \hline 5 & 7 & 1 & 1 & 4 \end{array}\right] = \begin{bmatrix} \overset{3}{B} & \overset{2}{C} \\ D & E \end{bmatrix}_{1}^{2} \tag{2.58}$$

where B, C, D, E are the arrays indicated by the dashed lines. The dimensions can be marked as shown in Eq. (2.58).

More generally, if some rows and columns of any matrix A are deleted then the resulting matrix is called a *submatrix* of A. For example, B in Eq. (2.58) is the submatrix obtained by deleting row 3 and columns 4 and 5 of A. It is a convention that A can be regarded as a submatrix of itself.

Example 2.15 (continued)   Deleting row 2 and columns 2, 4, 5 of the matrix A in Eq. (2.58) gives the following 2×2 submatrix

$$\begin{bmatrix} 1 & 2 \\ 5 & 1 \end{bmatrix}$$

When A is square, an important special type of submatrix is obtained by building up square arrays, starting in the top left corner, and finishing with A itself. These are called the *leading principal submatrices* of A.

Example 2.16 The leading principal submatrices of A in (2.51) are

$$[1], \quad \begin{bmatrix} 1 & 3 \\ 3 & 4 \end{bmatrix}, \quad A$$

which are built up as indicated below:

$$
\begin{array}{ccc}
1 & 3 & 7 \\
3 & 4 & 2 \\
7 & 2 & 0
\end{array}
\tag{2.59}
$$

It is clear from this example that the principal diagonal of each leading principal submatrix is part (or all) of the principal diagonal of A. More generally, any square submatrix of A whose principal diagonal satisfies this property is called simply a *principal submatrix* of A. Thus for A in (2.59), principal submatrices are

$$\begin{bmatrix} 4 & 2 \\ 2 & 0 \end{bmatrix}, \quad \begin{bmatrix} 1 & 7 \\ 7 & 0 \end{bmatrix} \tag{2.60}$$

Problem 2.28 What rows and columns of A in (2.59) are deleted to give (2.60)?

Partitioning is useful when applied to large matrices since manipulations can be carried out on the smaller blocks. For example, if A_1 is a 3×5 matrix partitioned into blocks B_1, C_1, D_1, E_1 in the same way as in (2.58) then

$$A + A_1 = \begin{bmatrix} (B + B_1) & (C + C_1) \\ (D + D_1) & (E + E_1) \end{bmatrix}$$

More importantly, when multiplying matrices in partitioned form the basic rule can be applied to the blocks as though they were single elements. For example, if

$$X = \begin{bmatrix} X_1 \\ X_2 \end{bmatrix} \begin{matrix} 3 \\ 2 \end{matrix} \tag{2.61}$$

then with A in (2.58) we obtain

$$AX = \begin{bmatrix} B & C \\ D & E \end{bmatrix}\begin{bmatrix} X_1 \\ X_2 \end{bmatrix}$$
$$= \begin{bmatrix} BX_1 + CX_2 \\ DX_1 + EX_2 \end{bmatrix} \tag{2.62}$$

The only restriction is that the blocks must be conformable for multiplication, so all the products BX_1, CX_2, etc., in (2.62) exist. This requires that in a product AX the number of *columns* in each block of A must equal the number of *rows* in the corresponding block of X (as illustrated in (2.58) and (2.61)).

Problem 2.29 Complete the partitioning for each of the following products so that each matrix is divided into four submatrices which are conformable for multiplication.

(a)

(b)

Problem 2.30 Choose arbitrary numbers for the elements of X_1 and X_2 in (2.61) and hence evaluate (2.62). Also evaluate AX directly without partitioning.

The rule for transposing a partitioned matrix is best explained by applying it to (2.58). We obtain

$$A^T = \begin{matrix} & 2 & 1 \\ & \end{matrix}\begin{bmatrix} B^T & D^T \\ C^T & E^T \end{bmatrix}\begin{matrix} 3 \\ 2 \end{matrix}$$

showing that the rows and columns of blocks are interchanged, and in addition each submatrix is itself transposed.

If A is square and its only nonzero elements can be partitioned as principal submatrices, then it is called *block diagonal*.

26

Example 2.17 The matrix

$$A = \begin{bmatrix} 1 & 2 & | & 0 \\ 3 & 4 & | & 0 \\ \hline 0 & 0 & | & 2 \end{bmatrix} \tag{2.63}$$

is block diagonal. A convenient notation which generalizes (2.36) is to write (2.63) as

$$A = \mathrm{diag}[A_1, A_2] \tag{2.64}$$

where

$$A_1 = \begin{bmatrix} 1 & 2 \\ 3 & 4 \end{bmatrix}, \qquad A_2 = [2]$$

Problem 2.31 If A_1 and A_2 in (2.64) are square matrices having arbitrary dimensions, obtain A^T and A^2 in partitioned form.

2.5 Kronecker product

The student should appreciate that we adopted the definition of multiplication in Section 2.2.3 because it arose in a natural way, and has properties which are useful in many applications. However, other definitions are possible, and as an important example of one of these we describe the *Kronecker* (or *direct*) product. Let A be the matrix in Eq. (2.2) and let B be another arbitrary matrix having dimensions $p \times q$. Then $A \otimes B$ is defined to be the partitioned matrix

$$A \otimes B = \begin{bmatrix} a_{11}B & a_{12}B & \cdots & a_{1n}B \\ \vdots & \vdots & & \vdots \\ a_{m1}B & a_{m2}B & \cdots & a_{mn}B \end{bmatrix} \tag{2.65}$$

Each submatrix in (2.65) has dimensions $p \times q$, so $A \otimes B$ has dimensions $(mp) \times (nq)$.

Example 2.18 Let A and B be the 2×2 matrices in (2.15). Then

$$A \otimes B = \begin{bmatrix} B & 2B \\ 3B & 4B \end{bmatrix} = \begin{bmatrix} -1 & 4 & -2 & 8 \\ 2 & -3 & 4 & -6 \\ -3 & 12 & -4 & 16 \\ 6 & -9 & 8 & -12 \end{bmatrix} \tag{2.66}$$

$$B \otimes A = \begin{bmatrix} -A & 4A \\ 2A & -3A \end{bmatrix} = \begin{bmatrix} -1 & -2 & 4 & 8 \\ -3 & -4 & 12 & 16 \\ 2 & 4 & -3 & -6 \\ 6 & 8 & -9 & -12 \end{bmatrix} \quad (2.67)$$

Notice that $A \otimes B \neq B \otimes A$, as for 'ordinary' multiplication. However, if in (2.67) the second and third columns are interchanged, and the second and third rows are interchanged, then (2.66) is obtained.

It can be shown in general that the elements of $B \otimes A$ are simply a rearrangement of those of $A \otimes B$, so $B \otimes A$ can be transformed into $A \otimes B$ by suitable row and column interchanges. Thus the difficulty of non-commutativity encountered with ordinary multiplication has to some extent been overcome. In other results also the Kronecker product has advantages, for example

$$(A \otimes B)^T = A^T \otimes B^T \quad (2.68)$$

so unlike (2.41) there is no reversal of order (see also Exercise 2.13). The Kronecker product is not just an interesting mathematical idea, as it has applications in statistics, numerical analysis, communications theory, etc.

Problem 2.32 Verify (2.68) by comparing the right-hand side with the transpose of (2.65). Similarly, show that $(A \otimes B)^* = A^* \otimes B^*$.

Problem 2.33 Show that the Kronecker product is distributive and associative, i.e.,

$$A \otimes (B + C) = A \otimes B + A \otimes C; \quad A \otimes (B \otimes C) = (A \otimes B) \otimes C.$$

2.6 Derivative of a matrix

If the elements $a_{ij}(t)$ of A are differentiable functions of t then the *derivative* of A, denoted by dA/dt or $\dot{A}$, is the matrix whose elements are da_{ij}/dt. The usual rules for differentiation still hold, for example:

$$\frac{d}{dt}(A + B) = \frac{dA}{dt} + \frac{dB}{dt} \quad (2.69)$$

$$\frac{d}{dt}(AB) = \left(\frac{dA}{dt}\right)B + A\left(\frac{dB}{dt}\right) \quad (2.70)$$

Note that when applying the 'product rule' (2.70) the order of the terms must be preserved, e.g., $\dot{A}B \neq B\dot{A}$.

Example 2.19 If

$$A = \begin{bmatrix} t^2 & t^3 \\ t & \sin t \end{bmatrix}, \quad B = \begin{bmatrix} e^{2t} & 0 \\ \cos t & 4 \end{bmatrix} \tag{2.71}$$

then

$$\frac{dA}{dt} = \begin{bmatrix} 2t & 3t^2 \\ 1 & \cos t \end{bmatrix}, \quad \frac{dB}{dt} = \begin{bmatrix} 2e^{2t} & 0 \\ -\sin t & 0 \end{bmatrix} \tag{2.72}$$

Problem 2.34 Evaluate $d(AB)/dt$ for the two matrices in (2.71): (a) by first determining AB; (b) by using (2.70).

Problem 2.35 Deduce, using A in (2.71), that in general $A\dot{A} \neq \dot{A}A$. Hence obtain expressions for $d(A^2)/dt$ and $d(A^3)/dt$ for an arbitrary square matrix A.

The *integral* of a matrix whose elements are integrable functions of t is defined similarly as the matrix obtained by integrating each element, i.e., the i,j element of $\int A \, dt$ is $\int a_{ij} \, dt$.

Exercises

2.1 Fill in the missing elements in the following product

$$\begin{bmatrix} 2 & 0 & 0 \\ \cdot & 2 & 0 \\ 0 & \cdot & 2 \end{bmatrix} \begin{bmatrix} \cdot & \cdot & 0 \\ 0 & \cdot & \cdot \\ 0 & 0 & 2 \end{bmatrix} = \begin{bmatrix} 4 & -2 & 0 \\ -2 & 5 & -2 \\ 0 & -2 & 5 \end{bmatrix}$$

2.2 The relationship between the input current i_1 and voltage v_1 and the output current i_2 and voltage v_2 for the four-terminal network in Fig. 2.1 is given by

$$\begin{bmatrix} v_1 \\ i_1 \end{bmatrix} = \begin{bmatrix} 1 + i\omega CR & R(2 + i\omega CR) \\ i\omega C & 1 + i\omega CR \end{bmatrix} \begin{bmatrix} v_2 \\ i_2 \end{bmatrix}$$

Find v_1 and i_1 in terms of v_3 and i_3 for the network shown in Fig. 2.2.

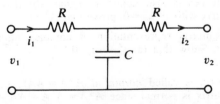

Fig. 2.1 Four-terminal network for Exercise 2.2.

29

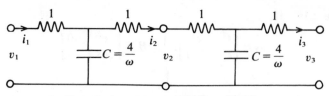

Fig. 2.2   Second network for Exercise 2.2.

2.3 Consider again the airlines problem described in Example 1.3 and Exercise 1.2. Verify that the product of the matrix of Eq. (1.1) and the matrix you found in part (a) of Exercise 1.2 is equal to the matrix of part (b), giving flight information from country A to country C.

2.4 For a square matrix, the diagonal from the top right corner to bottom left corner is called the *secondary* diagonal, and is perpendicular to the principal diagonal.

   Define J_n to be the $n \times n$ matrix having 1's along the secondary diagonal and zeros everywhere else, e.g.,

$$J_2 = \begin{bmatrix} 0 & 1 \\ 1 & 0 \end{bmatrix}$$

   (a) Prove $J_n^2 = I_n$. (b) If A is an arbitrary $n \times n$ matrix what are the products JA and AJ?

2.5 The generalization of the linear mapping in (1.7) can be written $x' = Tx$, where x and x' are column n-vectors and T is an $n \times n$ matrix, and represents a change from coordinates $x_1, \ldots, x_n$ to $x_1', \ldots, x_n'$. If $x_1' = 2x_1 + 3x_2$, $x_2' = 4x_3$, $x_3' = x_1 - 7x_2 + 3x_3$, write down T.

2.6 Use (2.41) to prove that $(A^T)^2 = (A^2)^T$ and similarly show that $(A^T)^k = (A^k)^T$ for any positive integer k and square matrix A.

2.7 The *trace* of an $n \times n$ matrix A is defined as the sum of the elements on the principal diagonal, i.e.,

$$\text{tr}(A) = a_{11} + a_{22} + \cdots + a_{nn} = \sum_{i=1}^{n} a_{ii} \tag{2.73}$$

Prove that: (a) $\text{tr}(A + B) = \text{tr}(A) + \text{tr}(B)$;

$$\text{(b) } \text{tr}(AB) = \text{tr}(BA); \text{ (c) } \text{tr}(AA^T) = \sum_{i=1}^{n} \sum_{j=1}^{n} a_{ij}^2$$

(use Eq. (2.19) for (b) and (c)).

2.8 A square matrix is called *upper triangular* if all the elements below the principal diagonal are zero. (a) Prove that the product of two upper triangular matrices is also upper triangular. (b) If A is a 3×3 upper triangular matrix with all $a_{ii} = 0$, prove that $A^3 = 0$.

2.9 Consider the matrix representation of complex numbers described in Example 2.10. Show that (a) $i^2 \sim -I_2$, (b) $\bar{z}_1 \sim (A^{(1)})^T$, where $A^{(1)}$ is the matrix in (2.33).

2.10 An $n \times n$ matrix is called *normal* if $A^*A = AA^*$. For example, a real symmetric matrix is normal since $A^*A = A^TA = AA = AA^*$. What other matrices are normal?

30

2.11 The *Fibonacci numbers* are $1, 1, 2, 3, 5, 8, 13, \ldots$, each number being obtained as the sum of the preceding two. These arise in many applications. If x_k denotes the kth number, then

$$x_{k+2} = x_{k+1} + x_k, \qquad k = 1, 2, 3, \ldots$$

with $x_1 = 1$, $x_2 = 1$. To obtain a matrix form define new variables $X_1(k) = x_k$, $X_2(k) = x_{k+1}$ so that $X_1(k + 1) = X_2(k)$ and $X_2(k + 1) = X_2(k) + X_1(k)$, and hence

$$\begin{bmatrix} X_1(k+1) \\ X_2(k+1) \end{bmatrix} = \begin{bmatrix} 0 & 1 \\ 1 & 1 \end{bmatrix} \begin{bmatrix} X_1(k) \\ X_2(k) \end{bmatrix} = A \begin{bmatrix} X_1(k) \\ X_2(k) \end{bmatrix} \tag{2.74}$$

$k = 1, 2, 3, \ldots$. The equations (2.74) represent a pair of linear *difference* equations. Show that

$$\begin{bmatrix} X_1(8) \\ X_2(8) \end{bmatrix} = A^7 \begin{bmatrix} 1 \\ 1 \end{bmatrix} = A^5 \begin{bmatrix} 2 \\ 3 \end{bmatrix}$$

and hence calculate x_8 and x_9.

2.12 If C, D are arbitrary matrices having dimensions $n \times r$ and $q \times s$ respectively, and A and B are as in (2.65), show that the *ordinary* product of $A \otimes B$ and $C \otimes D$ satisfies the relationship

$$(A \otimes B)(C \otimes D) = AC \otimes BD \tag{2.75}$$

by comparing terms on both sides of (2.75).

2.13 Define the *Kronecker power* of A by

$$A^{[2]} = A \otimes A, \ A^{[3]} = A \otimes A \otimes A = A \otimes A^{[2]}, \text{ etc.}$$

Use (2.75) to prove that $(AC)^{[2]}$ is equal to the ordinary product $A^{[2]}C^{[2]}$, and hence show that $(AC)^{[k]} = A^{[k]}C^{[k]}$ for any positive integer k. (A is $m \times n$, C is $n \times r$.)

Under what conditions on A and C does $(AC)^k = A^k C^k$?

2.14 The idea of Example 2.10 can be extended to the complex matrix A in (2.48), using the real partitioned matrix

$$D = \begin{bmatrix} A_1 & A_2 \\ -A_2 & A_1 \end{bmatrix}\begin{smallmatrix} n \\ n \end{smallmatrix} \sim A_1 + iA_2 = A \tag{2.76}$$

(a) Using the result of Problem 2.25a prove that if A is hermitian then D is symmetric.

(b) If $E \sim B_1 + iB_2 = B$, prove that $DE \sim AB$.

31

3. Unique solution of linear equations

In this chapter we shall study the solution of n simultaneous linear equations in n unknowns $x_1, x_2, \ldots, x_n$ in the form

$$\left.\begin{array}{c} a_{11}x_1 + a_{12}x_2 + \cdots + a_{1n}x_n = b_1 \\ a_{21}x_1 + a_{22}x_2 + \cdots + a_{2n}x_n = b_2 \\ \vdots \\ a_{n1}x_1 + a_{n2}x_2 + \cdots + a_{nn}x_n = b_n \end{array}\right\} \tag{3.1}$$

where the a's and b's are given real numbers. As illustrated in Example 1.4, Eqs (3.1) can be written

$$Ax = b \tag{3.2}$$

where $A = [a_{ij}]$ is the *matrix of coefficients* and

$$x = [x_1, x_2, \ldots, x_n]^T, \qquad b = [b_1, b_2, \ldots, b_n]^T \tag{3.3}$$

It is tempting to try to extend the ideas of matrix algebra developed in Chapter 2 so as to include 'division'. We could then solve (3.2) by writing $x = b \div A$, or borrowing the notation from ordinary algebra,

$$x = A^{-1}b \tag{3.4}$$

If we could find the matrix denoted by A^{-1} in (3.4), then this would give the desired solution vector x. However, it turns out that the best way to proceed is the reverse: we solve the equations, and then use this solution to find the matrix A^{-1}. The second part of this procedure will be delayed until Chapter 4. Notice incidentally that by applying the rule (2.41) to Eq. (3.2) these equations could equally well be written

$$x^T A^T = b^T$$

where both x^T and b^T are *row* vectors.

Equations with complex coefficients can be converted into the real case, as indicated in Exercise 3.7.

UNIQUE SOLUTION OF LINEAR EQUATIONS

3.1 Two equations and unknowns

Example 3.1 To solve the equations

$$x_1 + x_2 = 3 \tag{3.5}$$

$$2x_1 - 3x_2 = -4 \tag{3.6}$$

we use the familiar method of *elimination*: subtract twice Eq. (3.5) from Eq. (3.6) to obtain

$$-5x_2 = -10$$

so that $x_2 = 2$. Substitution of this value into (3.5) then gives $x_1 = 3 - 2 = 1$, so the equations have the unique solution $x_1 = 1$, $x_2 = 2$. In geometrical terms we can represent (3.5) and (3.6) as straight lines in the plane.

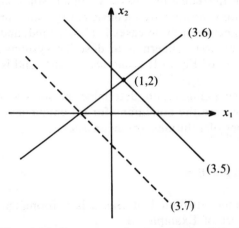

Fig. 3.1 The straight lines representing the equations (3.5), (3.6), (3.7).

The solution corresponds to the point of intersection of the lines (see Fig. 3.1).

Example 3.2 Suppose that (3.6) is replaced by

$$2x_1 + 2x_2 = -4 \tag{3.7}$$

Subtracting twice Eq. (3.5) from (3.7) gives $0 = -10$. This means that there are no values of x_1 and x_2 which satisfy (3.5) and (3.7) simultaneously. The two equations are called *inconsistent*. Geometrically, the two lines corresponding to (3.5) and (3.7) are parallel, and so have no point of intersection (see Fig. 3.1).

33

Example 3.3 Suppose instead that (3.6) is replaced by

$$2x_1 + 2x_2 = 6 \tag{3.8}$$

On subtracting twice (3.5) from (3.8) we get $0 = 0$, which doesn't seem to get us very far. What it means is that Eqs (3.5) and (3.8) do have a solution (they are *consistent*) but the second equation of the pair provides no information which is not given by the first. In other words, the solution of (3.5) and (3.8) is given by

$$x_1 = 3 - x_2 \tag{3.9}$$

where x_2 is arbitrary. Thus (3.9) represents an *infinite* number of solutions to the pair (3.5) and (3.8).

The three preceding examples illustrate what can happen in general for Eq. (3.1): either the equations have a unique solution, or an infinite number of solutions, or they are inconsistent (no solution). The problem of distinguishing between these cases is clearly vital, and will be tackled in Chapter 5. Our first concern is to describe systematic methods for finding the solution of Eq. (3.1) *assuming* it exists and is unique.

Problem 3.1 Using elimination, determine the solution of the Eqs (3.1) when $n = 2$. What are the conditions for (a) uniqueness of solution, (b) an infinite number of solutions, (c) inconsistency?

3.2 Gaussian elimination

A powerful method attributed to Gauss is a formalization of the elimination procedure of Example 3.1.

Example 3.4 To solve

$$x_1 - 3x_2 + 7x_3 = 2 \tag{3.10a}$$

$$2x_1 + 4x_2 - 3x_3 = -1 \tag{3.10b}$$

$$-3x_1 + 7x_2 + 2x_3 = 3 \tag{3.10c}$$

first eliminate x_1 from (3.10b) and (3.10c) by subtracting respectively 2 and -3 times (3.10a), to obtain

$$x_1 - 3x_2 + 7x_3 = 2 \tag{3.11a}$$

$$10x_2 - 17x_3 = -5 \tag{3.11b}$$

$$-2x_2 + 23x_3 = 9 \tag{3.11c}$$

Next eliminate x_2 from (3.11c) by multiplying (3.11b) by $-2/10$ and subtracting the result from (3.11c), so that the final set of equations becomes

$$x_1 - 3x_2 + 7x_3 = 2 \tag{3.12a}$$

$$10x_2 - 17x_3 = -5 \tag{3.12b}$$

$$\frac{196}{10} x_3 = 8 \tag{3.12c}$$

From (3.12c) $x_3 = 20/49$, and substituting into (3.12b) gives

$$x_2 = -\frac{1}{2} + \frac{17}{10}\left(\frac{20}{49}\right) = \frac{19}{98}$$

Finally, substituting into (3.12a)

$$x_1 = 2 + 3\left(\frac{19}{98}\right) - 7\left(\frac{20}{49}\right) = -\frac{27}{98}$$

so the solution is $(-27/98, 19/98, 20/49)$. The procedure for solving the Eqs (3.12) is called *back substitution*, since the unknowns are found in the order x_3, x_2, x_1.

It saves writing to record only the coefficients in the equations at each step of the process, and this can be done in matrix form. For (3.10) we have the *augmented matrix*

$$B = \begin{bmatrix} 1 & -3 & 7 & \vdots & 2 \\ 2 & 4 & -3 & \vdots & -1 \\ -3 & 7 & 2 & \vdots & 3 \end{bmatrix} \tag{3.13}$$

which consists of A together with the right-hand side of the equations as a fourth column, so in general for (3.1), B is the $n \times (n+1)$ matrix

$$B = [A, b] \tag{3.14}$$

In the solution of (3.10), at the first step which produces the Eqs (3.11) the first column of B is reduced so that it has all zeros below the 1,1 position, i.e.

$$B \rightarrow \begin{bmatrix} 1 & -3 & 7 & \vdots & 2 \\ 0 & 10 & -17 & \vdots & -5 \\ 0 & -2 & 23 & \vdots & 9 \end{bmatrix} \quad \begin{array}{l} (R2) - 2(R1) \\ (R3) - (-3)(R1) \end{array} \tag{3.15}$$

where the notation $(R2) - 2(R1)$ means the second row of B in (3.13) has twice the first row subtracted from it, and so on. The convention adopted is that the row which is altered is written *first*. Similarly the next step, corresponding to production of (3.12) is

$$\rightarrow \begin{bmatrix} 1 & -3 & 7 & \vdots & 2 \\ 0 & 10 & -17 & \vdots & -5 \\ 0 & 0 & \frac{196}{10} & \vdots & 8 \end{bmatrix} \quad (R3) - (-2/10)(R2) \tag{3.16}$$

35

Notice that the submatrix composed of the first three columns of the matrix in (3.16) is in *upper triangular* form – all the elements below the principal diagonal are zero (see Exercise 2.8). The corresponding equations (3.12) are called a *triangular* set, and as we have seen are easily solved.

Problem 3.2 Solve the equations

$$x_1 - x_2 + 3x_3 = 5$$
$$2x_1 - 4x_2 + 7x_3 = 7$$
$$4x_1 - 9x_2 + 2x_3 = -15$$

by gaussian elimination.

For $n > 3$, the method proceeds in exactly the same way. The first n columns of B are reduced one at a time, working from left to right, so as to obtain zeros below the principal diagonal. This is done by subtracting suitable multiples of rows. The resulting triangular system of equations is then solved by back substitution, starting with x_n.

In (3.15) the 1,1 element (equal to 1) is used to reduce the first column; in (3.16) the 2,2 element (equal to 10) is used to reduce the second column. These numbers are called the *pivots*, from the concept that the matrix changes or 'pivots' around these numbers. Generally, the ith pivot is the number in the (i,i) position which is used to reduce the ith column.

Suppose that columns $1, 2, \ldots, (j-1)$ of B have been reduced, and denote the elements in column j of B at this stage by b_{ij}, $i = 1, 2, \ldots, n$. The element b_{jj} on the principal diagonal cannot be used as pivot if it is zero. In this case the jth row must be interchanged with some row *below* it, say the kth, having $b_{kj} \neq 0$, before the elimination can be continued. This simply corresponds to interchanging the jth and kth equations, so does not affect the solution. In fact, if b_{jj} is nonzero but small compared with the other elements, then use of b_{jj} as pivot can lead to large rounding errors.

One rule which is often used to overcome both of these difficulties is to choose as pivot the element b_{kj} having largest numerical value in column j, i.e.,

$$|b_{kj}| = \max_i |b_{ij}|, \qquad j \leqslant i \leqslant n \tag{3.17}$$

Rows j and k are then interchanged and the next elimination step is carried out. This procedure is known as *partial pivoting*, and is done at *every* step of the process. Notice that in (3.17), rows above the jth are not considered since this would upset the triangular form already obtained for the first $j-1$ columns.

Example 3.5 Suppose at the third stage of a gaussian elimination with $n = 5$ the reduced matrix B is

$$\begin{bmatrix} 2 & 5 & 8 & \times & \times & \times \\ 0 & 3 & 1 & \times & \times & \times \\ 0 & 0 & 3 & \times & \times & \times \\ 0 & 0 & -7 & \times & \times & \times \\ 0 & 0 & 5 & \times & \times & \times \end{bmatrix} \qquad (3.18)$$

Using (3.17) the element in the 4,3 position (i.e., -7) would be used as the third pivot. The third and fourth rows in (3.18) would be interchanged, and the sub-diagonal elements in the third column would then be reduced to zero in the usual way.

The term 'partial' pivoting is used to contrast it with another scheme called 'complete pivoting'. In this the pivot is taken to be the element having largest numerical value in the lower right corner of the matrix – for example, in (3.18) this would be the array within dashed lines. This would involve both a row and a column interchange. In practice, however, partial pivoting is usually adequate.

If at stage j *all* the potential pivots b_{ij}, $j \leqslant i \leqslant n$, in column j are zero then the gaussian elimination procedure as described above stops. In fact this means that the equations do not have a unique solution. However, the method can be modified to determine whether the equations are consistent, and if so to calculate the (non-unique) solution. This will be discussed in detail in Chapter 5.

Problem 3.3 Solve the following set of equations by gaussian elimination, using partial pivoting and exact arithmetic.

$$5x_1 + x_2 + 2x_3 = 29$$
$$3x_1 - x_2 + x_3 = 10$$
$$x_1 + 2x_2 + 4x_3 = 31$$

Problem 3.4 Carry out gaussian elimination on the following set of equations, and hence deduce that they are inconsistent.

$$x_1 - 2x_2 + x_3 - x_4 = -5$$
$$x_1 + 5x_2 - 7x_3 + 2x_4 = 2$$
$$3x_1 + x_2 - 5x_3 + 3x_4 = 1$$
$$2x_1 + 3x_2 - 6x_3 = 21$$

When performing numerical calculations by hand it is useful to have checks against arithmetical errors. One simple check is to calculate the sums of the elements in each row of the original matrix B in (3.14), and regard these sums as forming an extra column. The elimination operations are then carried out in the usual way, including this new column. At each step the sum of the elements in each row of the reduced matrix B should equal the element in the same row of this extra column, if no slip has been made in the elimination procedure. A check on the back substitution process can be made by inserting the computed values of the variables into the original equations, which should be satisfied within the accuracy used.

Example 3.6 In Eq. (3.13) the row sums are 7, 2, 9 respectively. We have, using the operations in (3.15) and (3.16),

$$\begin{bmatrix} 7 \\ 2 \\ 9 \end{bmatrix} \rightarrow \begin{bmatrix} 7 \\ -12 \\ 30 \end{bmatrix} \quad \begin{array}{l} (R2) - 2(R1) \\ (R3) - (-3)(R1) \end{array} \tag{3.19}$$

$$\rightarrow \begin{bmatrix} 7 \\ -12 \\ 27\frac{3}{5} \end{bmatrix} \quad (R3) - (-2/10)(R2) \tag{3.20}$$

The elements in (3.19) and (3.20) agree with the row sums in (3.15) and (3.16) respectively.

Problem 3.5 Carry out the check on your working for Problems 3.2 and 3.3.

Problem 3.6 Use gaussian elimination to solve the equations

$$\begin{array}{rcl} x_1 + x_2 + x_3 + x_4 &=& -1 \\ 2x_1 - x_2 + 3x_3 &=& 1 \\ 2x_2 + 3x_4 &=& -1 \\ -x_1 + 2x_3 + x_4 &=& -2 \end{array}$$

3.3 Triangular decomposition

We have seen that triangular systems of equations can be solved very easily. This leads to the idea of expressing A in Eq. (3.2) as a product

$$A = LU \tag{3.21}$$

where $U = [u_{ij}]$ is an $n \times n$ upper triangular matrix (defined in Exercise 2.8), and $L = [l_{ij}]$ is an $n \times n$ *lower* triangular matrix (i.e., all elements of L above the principal diagonal are zero). It is convenient to take all $l_{ii} = 1$.

Example 3.7 Let A be the matrix of coefficients for the equations (3.10). Since $n = 3$ we have

$$LU = \begin{bmatrix} 1 & 0 & 0 \\ l_{21} & 1 & 0 \\ l_{31} & l_{32} & 1 \end{bmatrix} \begin{bmatrix} u_{11} & u_{12} & u_{13} \\ 0 & u_{22} & u_{23} \\ 0 & 0 & u_{33} \end{bmatrix} \tag{3.22}$$

$$= \begin{bmatrix} u_{11} & u_{12} & u_{13} \\ l_{21}u_{11} & (l_{21}u_{12} + u_{22}) & (l_{21}u_{13} + u_{23}) \\ l_{31}u_{11} & (l_{31}u_{12} + l_{32}u_{22}) & (l_{31}u_{13} + l_{32}u_{23} + u_{33}) \end{bmatrix} \tag{3.23}$$

The nine unknown elements of L and U can be obtained by equating the nine elements of (3.23) to those of the 3×3 matrix A. This is done in the following order:

First row of A: $\quad\quad u_{11} = 1, \quad\quad u_{12} = -3, \quad\quad u_{13} = 7 \quad\quad\quad$ (3.24)

First column of A: $\quad l_{21}u_{11} = 2, \quad\quad l_{31}u_{11} = -3$, whence

$$l_{21} = 2, \quad\quad l_{31} = -3 \tag{3.25}$$

Second row of A: $\quad l_{21}u_{12} + u_{22} = 4, \quad\quad l_{21}u_{13} + u_{23} = -3$, whence

$$u_{22} = 10, \quad\quad u_{23} = -17 \tag{3.26}$$

Second column of A: $\quad l_{31}u_{12} + l_{32}u_{22} = 7$, whence

$$l_{32} = -1/5 \tag{3.27}$$

Third row of A: $\quad\quad l_{31}u_{13} + l_{32}u_{23} + u_{33} = 2$, whence

$$u_{33} = 98/5 \tag{3.28}$$

Notice the order in which the elements of U and L are obtained in (3.24)–(3.28): first row of U, first column of L, second row of U, second column of L, third row of U. Substituting for the elements in (3.22) gives

$$A = \begin{bmatrix} 1 & 0 & 0 \\ 2 & 1 & 0 \\ -3 & -\dfrac{1}{5} & 1 \end{bmatrix} \begin{bmatrix} 1 & -3 & 7 \\ 0 & 10 & -17 \\ 0 & 0 & \dfrac{98}{5} \end{bmatrix} \tag{3.29}$$
$$\quad\quad\quad\quad\quad L \quad\quad\quad\quad\quad\quad U$$

We now show how the decomposition is used to solve Eq. (3.2). Substituting (3.21) into (3.2) gives

$$LUx = b \tag{3.30}$$

and this is written as *two* triangular sets of equations:

$$Ly = b, \quad\quad Ux = y \tag{3.31}$$

where $y = [y_1, y_2, \ldots, y_n]^T$. The first set is solved for y, and then the second set for x, which is the desired solution of the original equations.

Example 3.7 (continued)

Using (3.29) the first set of equations in (3.31) is

$$y_1 \qquad = 2$$
$$2y_1 + y_2 \qquad = -1$$
$$-3y_1 - \frac{1}{5}y_2 + y_3 = 3$$

(the b_i are as in (3.10)). The solution is therefore

$$y_1 = 2, \qquad y_2 = -1 - 2.2 = -5, \qquad y_3 = 3 + 3.2 + \frac{1}{5}(-5) = 8$$

The second set in (3.31) is

$$x_1 - 3x_2 + 7x_3 = 2$$
$$10x_2 - 17x_3 = -5$$
$$\frac{98}{5}x_3 = 8$$

so $x_3 = 20/49$, $x_2 = (-5 + 17x_3)/10 = 19/98$, $x_1 = 2 + 3x_2 - 7x_3 = -27/98$
which agrees, of course, with the values found in Example 3.4.

Problem 3.7 Determine L and U for the matrix

$$A = \begin{bmatrix} 2 & 3 & 4 \\ 4 & 10 & 9 \\ 6 & 17 & 20 \end{bmatrix}$$

and hence solve Eq. (3.2) in this case with $b = [23, 59, 101]^T$.

We now show how the rather tedious arithmetic of (3.24)–(3.28) can be avoided. Consider the decomposition of A in (3.29), and compare this with the gaussian reduction of the same matrix A in (3.16). It will be seen that U is identical to the triangularized form of A. In fact this is no coincidence, and furthermore L can also be obtained from the gaussian elimination procedure, as follows. Continuing with the same example, consider the step (3.15): to obtain zero in the 2,1 position we subtract 2 times (R1) from (R2), and for a zero in the 3,1 position, -3 times (R1) is subtracted from (R3). It is convenient to denote by m_{ij} the *multiplier* which is used to obtain zero in the i,j position, corresponding to the operation $(Ri) - m_{ij}(Rj)$. Thus in our example $m_{21} = 2$, $m_{31} = -3$, and similarly from (3.16) we see that $m_{32} = -2/10 = -1/5$. If we now write down a lower triangular matrix having 1's on the principal diagonal, and

40

m_{ij} as the i,j element, i.e.,

$$\begin{bmatrix} 1 & 0 & 0 \\ 2 & 1 & 0 \\ -3 & -\dfrac{1}{5} & 1 \end{bmatrix}$$

we see that this is identical to L in the decomposition (3.29).

It can be shown that the above method for obtaining $A = LU$ holds in general: If A is reduced to triangular form by gaussian elimination (without row interchanges), then the resulting upper triangular matrix is U, and the matrix of multipliers constructed as described above is L. If partial pivoting is used, then we get triangular factors such that $LU = A'$, where A' is the matrix which is obtained from the original A by applying the row interchanges to it in the same order. Thus gaussian elimination and LU decomposition are equivalent procedures, and involve the same total computational effort.

Problem 3.8 Repeat the decomposition for the matrix A in Problem 3.7, using gaussian elimination.

Problem 3.9 Solve the equations

$$2x_1 + x_2 + 2x_3 = 10$$
$$4x_1 + 4x_2 + 7x_3 = 33$$
$$2x_1 + 5x_2 + 12x_3 = 48$$

using LU decomposition. Hence find the solution if the right-hand sides of the equations change to 9, 22, 25 respectively.

An alternative way of performing the LU decomposition provides an interesting application of partitioning. Recall (Section 2.4) that the $r \times r$ leading principal submatrix A_r of A is the submatrix formed by rows and columns numbers $1, 2, \ldots, r$ of A. Suppose that

$$A_r = L_r U_r, \qquad A_{r+1} = L_{r+1} U_{r+1} \tag{3.32}$$

with L_r, L_{r+1} lower triangular and U_r, U_{r+1} upper triangular. Some manipulation shows that once L_r and U_r have been found then

$$L_{r+1} = \begin{bmatrix} L_r & 0 \\ c_r & 1 \end{bmatrix} \begin{matrix} r \\ 1 \end{matrix}, \qquad U_{r+1} = \begin{bmatrix} U_r & d_r \\ 0 & \alpha_r \end{bmatrix} \begin{matrix} r \\ 1 \end{matrix} \tag{3.33}$$

where the row r-vector c_r is given by

$$c_r U_r = [a_{r+1,1}, a_{r+1,2}, \ldots, a_{r+1,r}] \tag{3.34}$$

the column r-vector d_r by

$$L_r d_r = [a_{1,r+1}, a_{2,r+1}, \ldots, a_{r,r+1}]^T \tag{3.35}$$

and the scalar α_r by

$$\alpha_r = a_{r+1,r+1} - c_r d_r \tag{3.36}$$

The procedure is thus a repetitive one: L_1 and U_1 are both scalars so $L_1 = 1$, $U_1 = a_{11}$; (3.34) and (3.35) are solved for c_1 and d_1, and α_1 is found from (3.36), giving L_2 and U_2 in (3.33); the process is repeated with $r = 2$ to find L_3 and U_3, and so on, until $L_n = L$ and $U_n = U$ are determined.

Example 3.8 We repeat the decomposition of Example 3.7, using the matrix A of Eqs (3.10). Since $A_1 = 1$ we have $L_1 = 1$, $U_1 = 1$. Next,

$$A_2 = \begin{bmatrix} 1 & -3 \\ 2 & 4 \end{bmatrix}$$

so with $r = 1$ in (3.34) and (3.35) these become

$$c_1 \cdot 1 = a_{21} = 2, \qquad 1 \cdot d_1 = a_{12} = -3$$

and in (3.36)

$$\alpha_1 = a_{22} - c_1 d_1 = 4 + 6 = 10$$

Thus from (3.33) we obtain

$$L_2 = \begin{bmatrix} 1 & 0 \\ 2 & 1 \end{bmatrix}, \qquad U_2 = \begin{bmatrix} 1 & -3 \\ 0 & 10 \end{bmatrix} \tag{3.37}$$

The process is repeated with $r = 2$ in (3.34), to give

$$[c_{21}, c_{22}] U_2 = [a_{31}, a_{32}] = [-3, 7] \tag{3.38}$$

where U_2 is the matrix in (3.37). The equations in (3.38) are easily solved because U_2 is triangular:

$$c_{21} = -3, \qquad c_{22} = (7 + 3c_{21})/10 = -1/5$$

Similarly, from (3.35)

$$L_2 \begin{bmatrix} d_{21} \\ d_{22} \end{bmatrix} = \begin{bmatrix} a_{13} \\ a_{23} \end{bmatrix} = \begin{bmatrix} 7 \\ -3 \end{bmatrix}$$

whence

$$d_{21} = 7, \qquad d_{22} = -17$$

and finally, from (3.36)

$$\alpha_2 = a_{33} - c_2 d_2 = 2 - \begin{bmatrix} -3, & -\dfrac{1}{5} \end{bmatrix} \begin{bmatrix} 7 \\ -17 \end{bmatrix} = 2 - 21 + \frac{17}{5} = \frac{98}{5}$$

The desired L and U are then obtained by setting $r = 2$ in (3.33):

$$L_3 = \begin{bmatrix} L_2 & 0 \\ c_{21} \ c_{22} & 1 \end{bmatrix} \qquad U_3 = \begin{bmatrix} U_2 & \begin{matrix} d_{21} \\ d_{22} \end{matrix} \\ 0 & \alpha_2 \end{bmatrix}$$

and it can be seen that these agree with L and U found earlier in (3.29).

Problem 3.10 Repeat the determination of L and U for the matrix A in Problem 3.7 using the iterative process described above.

A word of warning is necessary here: not every matrix can be directly decomposed into factors L and U, as the following simple example demonstrates.

Example 3.9 For the equations

$$4x_2 = 8$$
$$4x_1 + 2x_2 = 17 \tag{3.39}$$

if we write

$$A = \begin{bmatrix} 0 & 4 \\ 4 & 2 \end{bmatrix} = \begin{bmatrix} 1 & 0 \\ l_1 & 1 \end{bmatrix} \begin{bmatrix} u_1 & u_2 \\ 0 & u_3 \end{bmatrix}$$

comparison of the 1,1 and 2,1 elements gives $u_1 = 0$, $l_1 u_1 = 4$, which cannot be satisfied. However, if the order of the equations in (3.39) is reversed then

$$A = \begin{bmatrix} 4 & 2 \\ 0 & 4 \end{bmatrix} = \underset{L}{\begin{bmatrix} 1 & 0 \\ 0 & 1 \end{bmatrix}} \underset{U}{\begin{bmatrix} 4 & 2 \\ 0 & 4 \end{bmatrix}}$$

This example shows that it may be necessary to alter the order of the equations before the decomposition (3.21) can be found. This is to ensure that no u_{rr} (or α_r in the partitioned form) is zero. It can be shown that this rearrangement is always possible provided the original equations (3.2) have a unique solution. In fact it can also be shown that the reordering must be carried out so that the new leading principal submatrices A_r have nonzero determinants (see Exercise 4.22). One important special case when A can be decomposed directly is when A is symmetric and positive definite (the latter is defined in Section 7.4) in which case L can be set equal to U^T (see Exercise 3.6).

Problem 3.11 Express the symmetric matrix

$$A = \begin{bmatrix} 4 & 1 & 3 \\ 1 & \dfrac{5}{2} & 0 \\ 3 & 0 & 15 \end{bmatrix}$$

in the form $U^T U$ where U is upper triangular.

3.4 Ill-conditioning

It should not be thought that all problems associated with solution of linear equations have now been overcome. The following example

43

provides an introduction to the sort of numerical complications which can occur.

Example 3.10 Consider the equations

$$28x_1 + 25x_2 = 30 \tag{3.40}$$

$$19x_1 + 17x_2 = 20 \tag{3.41}$$

and suppose that by some means we have calculated (to two significant figures) a solution $x_1 = 18$, $x_2 = -19$. When these values are substituted into the left-hand sides of (3.40) and (3.41) they give 29 and 19 respectively, so it would seem reasonable to assume that this solution is a close approximation to the correct one. In fact the *exact* solution of (3.40) and (3.41) is $x_1 = 10$, $x_2 = -10$.

Suppose next that the right-hand side of (3.41) changes from 20 to 19, all else remaining unaltered. Then the exact solution changes to $x_1 = 35$, $x_2 = -38$, a disproportionately large variation.

Equations (3.40) and (3.41) are an example of an ill-conditioned set, when small changes in coefficients in the equations lead to much larger changes in the values of the x's. Such systems need special attention. One sign of ill-conditioning is the presence of a relatively small pivot. A simple (although not foolproof) test for detecting ill-conditioning is to determine by recalculation whether the solution alters drastically after small changes have been made to the coefficients in the original problem.

We shall consider another kind of solution method in Section 6.7.1. However, further discussion of practical difficulties and methods which have been developed to overcome them lies outside the scope of this book, and the reader is referred to Fox (1964), Goult *et al.* (1974), and Stewart (1973). Such details are of primary interest to numerical specialists, since in most cases where large systems of linear equations have to be solved there will usually be access to a digital computer and a well tried and tested library program.

Problem 3.12 Using a pocket calculator, determine the solution of the equations

$$1.985x_1 - 1.358x_2 = 2.212$$
$$0.953x_1 - 0.652x_2 = b_2$$

(a) when $b_2 = 1.062$, (b) when $b_2 = 1.063$, giving your answers correct to three decimal places.

UNIQUE SOLUTION OF LINEAR EQUATIONS

Exercises

3.1 Using a pocket calculator, determine the solution of

$$3.41x_1 + 1.23x_2 - 1.09x_3 = 4.72$$
$$2.71x_1 + 2.14x_2 + 1.29x_3 = 3.10$$
$$1.89x_1 - 1.91x_2 - 1.89x_3 = 2.91$$

correct to two decimal places, using gaussian elimination with partial pivoting, and applying checks.

3.2 A square matrix is called *tridiagonal* if its only nonzero elements lie on the principal diagonal and on the diagonals immediately above and below this, e.g.,

$$A = \begin{bmatrix} a_1 & b_1 & 0 & 0 \\ c_1 & a_2 & b_2 & 0 \\ 0 & c_2 & a_3 & b_3 \\ 0 & 0 & c_3 & a_4 \end{bmatrix} \tag{3.42}$$

Express A as the product LU in the form

$$A = \begin{bmatrix} 1 & 0 & 0 & 0 \\ l_1 & 1 & 0 & 0 \\ 0 & l_2 & 1 & 0 \\ 0 & 0 & l_3 & 1 \end{bmatrix} \begin{bmatrix} u_1 & v_1 & 0 & 0 \\ 0 & u_2 & v_2 & 0 \\ 0 & 0 & u_3 & v_3 \\ 0 & 0 & 0 & u_4 \end{bmatrix} \tag{3.43}$$

where the u's are all nonzero, and determine the conditions on the a's, b's, c's for (3.43) to hold. This can be extended to $n \times n$ tridiagonal matrices with L, U bidiagonal, as in (3.43), and gives a simple method for solving Eq. (3.2) in this case.

3.3 Using Eq. (3.43), solve the equations

$$\begin{aligned} x_1 + 2x_2 & & & = 8 \\ 2x_1 - x_2 + x_3 & & & = -1 \\ 3x_2 - 3x_3 - x_4 & & & = 10 \\ 7x_3 + 4x_4 & & & = 6 \end{aligned}$$

3.4 The equation of a circle in cartesian coordinates is $x_1^2 + x_2^2 + 2gx_1 + 2fx_2 + c = 0$. Determine the equation of the circle which passes through the points $(7, 5)$, $(6, -2)$, $(-1, -1)$, by solving the appropriate equations using gaussian elimination.

3.5 Use gaussian elimination to find the matrix X such that $AX = B$, where

$$A = \begin{bmatrix} 4 & 2 & -1 \\ 5 & 3 & -1 \\ 3 & -1 & 4 \end{bmatrix}, \quad B = \begin{bmatrix} 1 & 3 \\ -1 & 1 \\ 6 & 4 \end{bmatrix}$$

3.6 Show that a real symmetric 2×2 matrix A can be expressed as $U^T U$, with U a real upper triangular matrix having nonzero diagonal elements, only if $a_{11} > 0$, $a_{11}a_{22} > a_{12}^2$.
(See Exercise 7.2 for a generalization of this result.)

3.7 Systems of equations with complex coefficients can be converted into the real case as follows:
Let $A = A_1 - iA_2$, $b = b_1 + ib_2$, $x = x_1 + ix_2$ in Eq. (3.2) with the matrices

45

A_1, A_2, and the column vectors b_1, b_2, x_1, x_2 all real. Hence show that Eq. (3.2) can be written in the real form

$$D\begin{bmatrix} x_1 \\ x_2 \end{bmatrix} = \begin{bmatrix} b_1 \\ b_2 \end{bmatrix}$$

where D is the $2n \times 2n$ real matrix defined in (2.76).

3.8 Write a computer program to solve four equations in four unknowns using gaussian elimination. Incorporate the possibility of row interchanges to deal with zero pivots, but do not include partial pivoting. Test your program on Problem 3.6.

Use your program to solve the system of equations for which

$$[A, b] = \begin{bmatrix} 0.21145 & 2.29634 & 2.71542 & 3.21468 & 8.43789 \\ 0.43718 & 3.91569 & 1.68260 & 2.85212 & 8.88759 \\ 6.09889 & 4.32403 & 23.2021 & 1.57816 & 35.20318 \\ 4.62304 & 0.89257 & 15.3216 & 5.30477 & 26.14198 \end{bmatrix}$$

Compare your results with the exact answer $x_1 = x_2 = x_3 = x_4 = 1$. Rewrite your program to include partial pivoting, and recompute a solution.

4. Determinant and inverse

The reader who worked through Problem 3.1 will have found that the solution of

$$a_{11}x_1 + a_{12}x_2 = b_1$$
$$a_{21}x_1 + a_{22}x_2 = b_2 \tag{4.1}$$

is

$$x_1 = (a_{22}b_1 - a_{12}b_2)/d, \qquad x_2 = (-a_{21}b_1 + a_{11}b_2)/d \tag{4.2}$$

where

$$d = a_{11}a_{22} - a_{12}a_{21} \tag{4.3}$$

The solution (4.2) is valid only if $d \neq 0$, so d *determines* whether the equations (4.1) have a unique solution. For this reason d is called the *determinant* of the matrix of coefficients

$$A = \begin{bmatrix} a_{11} & a_{12} \\ a_{21} & a_{22} \end{bmatrix} \tag{4.4}$$

and is written $\det A$, $|A|$,

$$\det \begin{bmatrix} a_{11} & a_{12} \\ a_{21} & a_{22} \end{bmatrix} \quad \text{or} \quad \begin{vmatrix} a_{11} & a_{12} \\ a_{21} & a_{22} \end{vmatrix}$$

We thus have the *definition*

$$\det A = a_{11}a_{22} - a_{12}a_{21} \tag{4.5}$$

for the general 2×2 matrix A in (4.4). In fact determinants were studied long before the introduction of matrices but are now much less important.

The solution (4.2) can be written as

$$\begin{bmatrix} x_1 \\ x_2 \end{bmatrix} = \frac{1}{d} \begin{bmatrix} a_{22} & -a_{12} \\ -a_{21} & a_{11} \end{bmatrix} \begin{bmatrix} b_1 \\ b_2 \end{bmatrix} \tag{4.6}$$

so by comparison with Eq. (3.4), i.e., $x = A^{-1}b$, we can identify in this case

$$A^{-1} = \frac{1}{d} \begin{bmatrix} a_{22} & -a_{12} \\ -a_{21} & a_{11} \end{bmatrix} \tag{4.7}$$

as the 'inverse' of A. It is left as an easy exercise to verify that for the

47

matrices in (4.4) and (4.7)

$$AA^{-1} = A^{-1}A = I_2 \tag{4.8}$$

which adds further justification for the notation A^{-1}.

The aim of this chapter is to develop properties of the determinant and inverse for general square matrices.

Problem 4.1 For the matrix $A^{(1)}$ in (2.33) associated with the complex number z_1, show (a) $\det A^{(1)} = |z_1|^2$, (b) $(A^{(1)})^{-1} \sim z_1^{-1}$.

4.1 Determinant

4.1.1 3×3 case

It is tedious but straightforward to solve Eqs (3.1) with $n = 3$, and it turns out, as in (4.2), that the expressions for x_1, x_2, x_3 have a common denominator which in this case is

$$d = a_{11}a_{22}a_{33} - a_{11}a_{23}a_{32} + a_{12}a_{23}a_{31} - a_{12}a_{21}a_{33} + a_{13}a_{21}a_{32} - a_{13}a_{22}a_{31} \tag{4.9}$$

Again, the equations have a unique solution if and only if (4.9) is nonzero. It is convenient to factorize (4.9) as follows:

$$d = a_{11}(a_{22}a_{33} - a_{23}a_{32}) - a_{12}(a_{21}a_{33} - a_{23}a_{31}) + a_{13}(a_{21}a_{32} - a_{22}a_{31}) \tag{4.10}$$

$$= a_{11}\begin{vmatrix} a_{22} & a_{23} \\ a_{32} & a_{33} \end{vmatrix} - a_{12}\begin{vmatrix} a_{21} & a_{23} \\ a_{31} & a_{33} \end{vmatrix} + a_{13}\begin{vmatrix} a_{21} & a_{22} \\ a_{31} & a_{32} \end{vmatrix} \tag{4.11}$$

the transition from (4.10) to (4.11) being accomplished using (4.5). The expression (4.11) can be regarded as the *definition* of the determinant of a general 3×3 matrix $A = [a_{ij}]$.

The notation introduced at the beginning of this chapter is used in general: for a square array the matrix is denoted by brackets and the determinant by vertical lines.

Problem 4.2 Using (4.11) calculate $\det A$ when

$$A = \begin{bmatrix} 2 & -2 & 5 \\ 1 & 7 & -2 \\ 4 & -3 & 6 \end{bmatrix}$$

Our definition of a 3×3 determinant in (4.11) is in terms of 2×2 determinants. To develop the definition of an $n \times n$ determinant depends upon appreciating how (4.11) is built up. The first 2×2 determinant in (4.11) is the determinant of the 2×2 submatrix obtained from A by deleting row 1 and column 1:

$$\begin{array}{ccc} \textcircled{a_{11}} & a_{12} & a_{13} \\ a_{21} & a_{22} & a_{23} \\ a_{31} & a_{32} & a_{33} \end{array} \tag{4.12}$$

To obtain the second term in (4.11) row 1 and column 2 are deleted:

$$\begin{array}{ccc} a_{11} & \textcircled{a_{12}} & a_{13} \\ a_{21} & a_{22} & a_{23} \\ a_{31} & a_{32} & a_{33} \end{array} \tag{4.13}$$

and similarly for the third term:

$$\begin{array}{ccc} a_{11} & a_{12} & \textcircled{a_{13}} \\ a_{21} & a_{22} & a_{23} \\ a_{31} & a_{32} & a_{33} \end{array} \tag{4.14}$$

It is convenient to define the *minor* M_{ij} of a_{ij} as the determinant of the submatrix obtained from A by deleting row i and column j. We can then write (4.12), (4.13), (4.14) respectively as

$$M_{11} = \begin{vmatrix} a_{22} & a_{23} \\ a_{32} & a_{33} \end{vmatrix}, \qquad M_{12} = \begin{vmatrix} a_{21} & a_{23} \\ a_{31} & a_{33} \end{vmatrix}, \qquad M_{13} = \begin{vmatrix} a_{21} & a_{22} \\ a_{31} & a_{32} \end{vmatrix} \tag{4.15}$$

and (4.11) then becomes

$$\det A = a_{11}M_{11} - a_{12}M_{12} + a_{13}M_{13} \tag{4.16}$$

The negative term in (4.16) can be avoided by defining the *cofactor* A_{ij} of a_{ij} as

$$A_{ij} = (-1)^{i+j}M_{ij} \tag{4.17}$$

so $A_{11} = (-1)^2 M_{11}$, $A_{12} = (-1)^3 M_{12}$, $A_{13} = (-1)^4 M_{13}$, and (4.16) becomes

$$\det A = a_{11}A_{11} + a_{12}A_{12} + a_{13}A_{13} \tag{4.18}$$

The formula (4.18) is called the *expansion* of $\det A$ by the first row, since it expresses $\det A$ as a term-by-term product of the elements in the first row with their cofactors.

Example 4.1 If

$$A = \begin{bmatrix} 1 & -3 & 7 \\ 2 & 4 & -3 \\ -3 & 7 & 2 \end{bmatrix} \tag{4.19}$$

then (4.18) gives

$$\det A = 1(-1)^2 \begin{vmatrix} 4 & -3 \\ 7 & 2 \end{vmatrix} + (-3)(-1)^3 \begin{vmatrix} 2 & -3 \\ -3 & 2 \end{vmatrix} + 7(-1)^4 \begin{vmatrix} 2 & 4 \\ -3 & 7 \end{vmatrix}$$
$$= (8 + 21) + 3(4 - 9) + 7(14 + 12)$$
$$= 196$$

49

An unexpected fact is that $\det A$ can be expanded by *any* row (or *any* column): form the term-by-term product of the elements in any row (or column) with their cofactors. For example, expanding by the third row

$$\det A = a_{31}A_{31} + a_{32}A_{32} + a_{33}A_{33}$$

and the reader can easily verify that this gives the same expression as (4.10).

Example 4.1 (continued)   Expanding $\det A$ by the second column gives

$$\det A = a_{12}A_{12} + a_{22}A_{22} + a_{32}A_{32} \tag{4.20}$$

$$= (-3)(-1)^3 \begin{vmatrix} 2 & -3 \\ -3 & 2 \end{vmatrix} + 4(-1)^4 \begin{vmatrix} 1 & 7 \\ -3 & 2 \end{vmatrix} + 7(-1)^5 \begin{vmatrix} 1 & 7 \\ 2 & -3 \end{vmatrix}$$

$$= 3(4-9) + 4(2+21) - 7(-3-14)$$

$$= 196, \text{ as before.}$$

Problem 4.3   Evaluate (4.20) for a general 3×3 matrix and confirm that it agrees with (4.10).

A second unexpected property is that if the elements in any row are multiplied term-by-term with the cofactors of a *different* row, the result is zero. For example, the first row taken with the cofactors of the third row gives

$$a_{11}A_{31} + a_{12}A_{32} + a_{13}A_{33} \tag{4.21}$$

$$= a_{11}(-1)^4 \begin{vmatrix} a_{12} & a_{13} \\ a_{22} & a_{23} \end{vmatrix} + a_{12}(-1)^5 \begin{vmatrix} a_{11} & a_{13} \\ a_{21} & a_{23} \end{vmatrix} + a_{13}(-1)^6 \begin{vmatrix} a_{11} & a_{12} \\ a_{21} & a_{22} \end{vmatrix}$$

$$= a_{11}(a_{12}a_{23} - a_{13}a_{22}) - a_{12}(a_{11}a_{23} - a_{13}a_{21}) + a_{13}(a_{11}a_{22} - a_{12}a_{21})$$

$$= 0$$

The same fact holds for columns: for example, taking the second column with the cofactors of the first column gives

$$a_{12}A_{11} + a_{22}A_{21} + a_{32}A_{31} = 0 \tag{4.22}$$

Problem 4.4   Verify (4.22) by evaluating in full.

4.1.2   General properties

The preceding results can be generalized for any $n \times n$ matrix A: we define the nth-order determinant by

$$\det A = a_{i1}A_{i1} + a_{i2}A_{i2} + \cdots + a_{in}A_{in} \quad \text{(expansion by ith row)} \tag{4.23}$$

$$= a_{1j}A_{1j} + a_{2j}A_{2j} + \cdots + a_{nj}A_{nj} \quad \text{(expansion by jth column)} \tag{4.24}$$

where A_{ij}, the cofactor of a_{ij}, is defined by (4.17). The minor M_{ij} is also defined as previously, being the determinant of the submatrix obtained by deleting row i, column j of A, but now has dimensions $(n-1) \times (n-1)$. Thus, for example, a 4×4 determinant is defined in terms of four 3×3 determinants, and so on. It is helpful to note that the sign $(-1)^{i+j}$ associated with A_{ij} can be taken from the pattern in Table 4.1, for example $A_{33} = + M_{33}$.

		j				
		1	2	3	4	$\cdots$
	1	+	−	+	−	$\cdots$
	2	−	+	−	+	$\cdots$
i	3	+	−	+	−	$\cdots$
	$\vdots$	$\vdots$				

Table 4.1

Problem 4.5 Evaluate the following by expanding by the first row

$$\begin{vmatrix} 1 & 0 & 2 & 0 \\ -1 & 4 & 3 & 6 \\ 0 & -2 & 5 & -3 \\ 3 & 1 & 1 & 0 \end{vmatrix}$$

Problem 4.6 Prove that if all the elements in a row (or column) are zero then $\det A = 0$.

The result on multiplying rows with cofactors of different rows still holds:

$$a_{i1}A_{k1} + a_{i2}A_{k2} + \cdots + a_{in}A_{kn} = 0, \qquad k \neq i \qquad (4.25)$$

and similarly for columns:

$$a_{1j}A_{1k} + a_{2j}A_{2k} + \cdots + a_{nj}A_{nk} = 0, \qquad k \neq j \qquad (4.26)$$

It must be stressed that, except for $n = 2$ or 3, or when many of the elements are zero, the rules (4.23) and (4.24) are in general too unwieldy for practical evaluation of determinants. This topic will be covered separately in Section 4.2. However, (4.23) and (4.24) are needed when the elements of A are algebraic quantities rather than numbers, and this is important for the determination of the characteristic equation, studied in Chapter 6. The formulae (4.23)–(4.26) are also valuable for theoretical purposes, and can be used to derive the following properties of $n \times n$ determinants.

Property PD1

$$\det A^T = \det A, \quad \det A^* = \overline{(\det A)} \tag{4.27}$$

Property PD2

$$\det(kA) = k^n \det A \tag{4.28}$$

where k is a scalar (recall that kA was defined in Section 2.2.2).

Instead of always saying 'rows or columns' it is useful to use the word 'line' to refer to both, with the understanding that when we mention two lines they are to be parallel.

Property PD3

If any two lines of A are identical then $\det A = 0$.

Example 4.2 The preceding three properties can be demonstrated for the 2×2 matrix in Eq. (4.4). First

$$A^T = \begin{bmatrix} a_{11} & a_{21} \\ a_{12} & a_{22} \end{bmatrix}, \qquad A^* = \begin{bmatrix} \bar{a}_{11} & \bar{a}_{21} \\ \bar{a}_{12} & \bar{a}_{22} \end{bmatrix}$$

so from (4.5)

$$\det A^T = a_{11}a_{22} - a_{21}a_{12} = \det A$$
$$\det(A^*) = \bar{a}_{11}\bar{a}_{22} - \bar{a}_{21}\bar{a}_{12} = \overline{(a_{11}a_{22} - a_{21}a_{12})} = \overline{(\det A)}$$

Next,

$$kA = \begin{bmatrix} ka_{11} & ka_{12} \\ ka_{21} & ka_{22} \end{bmatrix}$$

so from (4.5)

$$\det A = (ka_{11})(ka_{22}) - (ka_{12})(ka_{21})$$
$$= k^2 \det A$$

agreeing with (4.28). Finally, if the second column of A is identical to the first, i.e., $a_{12} = a_{11}$, $a_{22} = a_{21}$, then again from (4.5)

$$\det A = a_{11}a_{22} - a_{11}a_{22} = 0$$

showing that PD3 holds.

The properties can also be used to prove facts about determinants without consideration of elements, as the following example illustrates.

Example 4.3 If A is skew symmetric and n is an odd integer, then by definition $A^T = -A$ (see Section 2.3.2) so

$$\det A^T = \det(-A)$$
$$= (-1)^n \det A, \quad \text{by (4.28)}$$

Thus applying (4.27) we deduce that

$$\det A = (-1)^n \det A$$
$$= -\det A$$

since n is odd, whence $\det A = 0$ for any such matrix A.

Problem 4.7 Prove that the determinant of any hermitian matrix is a purely real number.

Property PD4
  If A is a diagonal matrix then

$$\det A = a_{11}a_{22}a_{33}\cdots a_{nn} \tag{4.29}$$

Property PD5
  If A is a triangular matrix then

$$\det A = a_{11}a_{22}a_{33}\cdots a_{nn} \tag{4.30}$$

Example 4.4 Using PD4 and PD5, we have

$$\begin{vmatrix} 2 & 0 & 0 \\ 0 & 4 & 0 \\ 0 & 0 & 7 \end{vmatrix} = 56, \qquad \begin{vmatrix} 2 & 3 & 5 \\ 0 & 4 & 1 \\ 0 & 0 & 7 \end{vmatrix} = 56$$

The simple result (4.30) for evaluation of triangular determinants is interesting because it suggests a link with gaussian elimination of Section 3.2, whereby a square matrix can be reduced to triangular form. This relationship will be developed in Section 4.2, but first we need some further important properties.

Property PD6
  (a) If any two lines of A are interchanged the value of $\det A$ is multiplied by -1.
  (b) If any line of A is multiplied by a nonzero scalar k then $\det A$ is also multiplied by k.
  (c) The value of $\det A$ is unchanged if an arbitrary multiple of any line is added to any other line.

Example 4.5 Let A be the 3×3 matrix in (4.19), which has $\det A = 196$. Then interchanging the first and third rows, denoted by $(R1) \leftrightarrow (R3)$, gives

$$\begin{vmatrix} -3 & 7 & 2 \\ 2 & 4 & -3 \\ 1 & -3 & 7 \end{vmatrix} = -196, \qquad \text{by PD6(a).}$$

Similarly, by PD6(c), $(R2) - 2(R1)$ and $(R3) + 3(R1)$ applied to (4.19) produce

$$\begin{vmatrix} 1 & -3 & 7 \\ 0 & 10 & -17 \\ 0 & -2 & 23 \end{vmatrix} = 196 \tag{4.31}$$

Problem 4.8 Prove that if one line of A is a multiple of another line, then $\det A = 0$.

The properties PD6 are also useful when the elements of a determinant are given as algebraic rather than numerical quantities, as the following two examples illustrate.

Example 4.6

$$\begin{vmatrix} 1 & 1 & 1 \\ a & b & c \\ a^2 & b^2 & c^2 \end{vmatrix} = \begin{vmatrix} 1 & 0 & 0 \\ a & (b-a) & (c-a) \\ a^2 & (b^2-a^2) & (c^2-a^2) \end{vmatrix} \qquad \begin{array}{l} (C2)-(C1) \\ (C3)-(C1) \end{array}$$

$$= \begin{vmatrix} (b-a) & (c-a) \\ (b^2-a^2) & (c^2-a^2) \end{vmatrix}, \qquad \text{expanding by first row}$$

$$= (b-a)(c-a)\begin{vmatrix} 1 & 1 \\ b+a & c+a \end{vmatrix} \tag{4.32}$$

$$= (b-a)(c-a)(c+a-b-a)$$

$$= (b-a)(c-a)(c-b)$$

Note that (4.32) is obtained from the preceding determinant by applying PD6(b) in reverse, so that the factors $(b-a)$ and $(c-a)$ can be removed from the first and second columns respectively. (Following Section 3.2, the notation $(C2)-(C1)$ means column 2 has column 1 subtracted from it, etc.)

Example 4.7

$$\begin{vmatrix} a & b+c & 1 \\ b & c+a & 1 \\ c & a+b & 1 \end{vmatrix} = \begin{vmatrix} a & a+b+c & 1 \\ b & a+b+c & 1 \\ c & a+b+c & 1 \end{vmatrix} \qquad (C2)+(C1)$$

$$= (a+b+c)\begin{vmatrix} a & 1 & 1 \\ b & 1 & 1 \\ c & 1 & 1 \end{vmatrix}, \qquad \text{by PD6(b)}$$

$$= 0, \qquad \text{by PD3.}$$

Problem 4.9 Express the 4×4 determinant in Problem 4.5 in terms of a single 3×3 determinant by reducing the first row to 1,0,0,0. Hence evaluate the determinant.

Problem 4.10 Without evaluating directly any of the determinants, prove

(a) $\begin{vmatrix} a-b & b-c & c-a \\ b-c & c-a & a-b \\ c-a & a-b & b-c \end{vmatrix} = 0$

(b) $\begin{vmatrix} (1+a) & b & c \\ a & (1+b) & c \\ a & b & (1+c) \end{vmatrix} = 1+a+b+c$

(c) $\begin{vmatrix} bcd & a & a^2 & a^3 \\ acd & b & b^2 & b^3 \\ abd & c & c^2 & c^3 \\ abc & d & d^2 & d^3 \end{vmatrix} = \begin{vmatrix} 1 & a^2 & a^3 & a^4 \\ 1 & b^2 & b^3 & b^4 \\ 1 & c^2 & c^3 & c^4 \\ 1 & d^2 & d^3 & d^4 \end{vmatrix}$

Property PD7

If A and B are two $n \times n$ matrices then

$$\det(AB) = (\det A)(\det B) \tag{4.33}$$

The result in (4.33) is by no means obvious: it states that if the product AB is evaluated then $\det(AB)$ is equal to the product of the determinants of A and of B. Note, however, that in general $\det(A + B) \neq \det A + \det B$ (try some simple examples).

Example 4.8 If A is decomposed into triangular factors as in (3.21), then

$$\begin{aligned} \det A &= \det(LU) \\ &= (\det L)(\det U), \qquad \text{by PD7} \\ &= (l_{11}l_{22} \ldots l_{nn})(u_{11}u_{22} \ldots u_{nn}), \qquad \text{by PD5} \\ &= u_{11}u_{22} \ldots u_{nn} \end{aligned}$$

since all $l_{ii} = 1$.

Problem 4.11 If A and B are two arbitrary $n \times n$ matrices, prove (a) $\det(AB) = \det(BA)$, (b) $\det(A^k) = (\det A)^k$ for any positive integer k.

To close this section it is worth noting that the term 'minor' is also used to denote the determinant of *any* square submatrix of A, not just one obtained by deleting only a single row and column. In particular the determinants of the (leading) principal submatrices of A (defined in Section 2.4) are called *(leading) principal minors* of A.

We have deliberately not given proofs of the properties of determinants presented in this section. Such proofs are relatively unimportant, except to mathematicians, and details can be found in very

55

many books (e.g., Mirsky (1963)). What *is* important for readers of this book is an appreciation of the properties and the ways in which they can be applied.

4.1.3 Some applications

It is somewhat fashionable to play down the usefulness of determinants. However, they still play a role in many problems of practical interest not always directly connected with sets of linear equations.

Example 4.9 In the algebra of vectors in three dimensions, the *vector product* of two vectors **a** and **b** is a vector perpendicular to both of them defined by

$$\mathbf{a} \times \mathbf{b} = \begin{vmatrix} \mathbf{i} & \mathbf{j} & \mathbf{k} \\ a_1 & a_2 & a_3 \\ b_1 & b_2 & b_3 \end{vmatrix} \tag{4.34}$$

where the determinant is to be expanded by the first row (expansion in any other way is not meaningful here). In (4.34) **i**, **j** and **k** are the unit vectors along the rectangular cartesian coordinate axes and $a_1, a_2, a_3, b_1, b_2, b_3$ are the components of **a** and **b** respectively in these directions. The *modulus* (or length) of any vector $\mathbf{x} = x_1\mathbf{i} + x_2\mathbf{j} + x_3\mathbf{k}$ is

$$|\mathbf{x}| = (x_1^2 + x_2^2 + x_3^2)^{1/2} \tag{4.35}$$

It can be shown that the area of a triangle having **a** and **b** as two sides is $\frac{1}{2}|\mathbf{a} \times \mathbf{b}|$.

Problem 4.12 If $\mathbf{a} = 2\mathbf{i} + 3\mathbf{j} - \mathbf{k}$, $\mathbf{b} = \mathbf{i} - \mathbf{j} + \mathbf{k}$, determine $\mathbf{a} \times \mathbf{b}$. Hence evaluate the area of the triangle having **a** and **b** as adjacent sides.

Problem 4.13 When **x** and **y** are vectors in three dimensions the *scalar product* in (2.44) is written

$$\mathbf{x} \cdot \mathbf{y} = x_1 y_1 + x_2 y_2 + x_3 y_3$$

It can be shown that the volume V of a tetrahedron having the vectors **a**, **b**, **c** as concurrent edges is given in terms of the *scalar triple product* $(\mathbf{a} \times \mathbf{b}) \cdot \mathbf{c}$, i.e.,

$$V = \frac{1}{6}(\mathbf{a} \times \mathbf{b}) \cdot \mathbf{c}$$

Use (4.34) to prove that

$$V = \frac{1}{6} \begin{vmatrix} a_1 & a_2 & a_3 \\ b_1 & b_2 & b_3 \\ c_1 & c_2 & c_3 \end{vmatrix}$$

Example 4.10 In control theory a system is said to be *controllable* if it is possible to manipulate the control variables in such a way that the system starts out from any initial state and finishes up in any desired state – for example, transferring a spacecraft from an orbit round the Earth to a specified orbit round the Moon, or to a 'soft landing' on the Moon, by suitably controlling the rocket motors. A large class of control systems can be described by a set of linear differential equations

$$\frac{dx}{dt} = Ax + bu \qquad (4.36)$$

where $x = [x_1, x_2, \ldots, x_n]^T$ is the vector of variables describing the state of the system, $dx/dt = [dx_1/dt, \ldots, dx_n/dt]^T$, A is a given constant $n \times n$ matrix, b is a given constant column n-vector, and u is the scalar control variable which can be manipulated.

It can be shown that (4.36) is controllable if and only if det $\mathscr{C} \neq 0$, where $\mathscr{C}$ is the *controllability* matrix and has columns $b, Ab, A^2b, \ldots, A^{n-1}b$, i.e.,

$$\mathscr{C} = [b, Ab, A^2b, \ldots, A^{n-1}b] \qquad (4.37)$$

Problem 4.14 It can be shown that a certain circuit, represented in Fig. 4.1, is described by (4.36) with

$$A = \begin{bmatrix} -\dfrac{1}{R_1C} & 0 \\ 0 & -\dfrac{R_2}{L} \end{bmatrix}, \qquad b = \begin{bmatrix} \dfrac{1}{R_1C} \\ \dfrac{1}{L} \end{bmatrix}$$

Obtain the condition which R_1, R_2, L and C must satisfy for the circuit to be controllable (i.e., it must be possible to change the voltage x_1 across the capacitor and the current x_2 through the inductor from any initial values to any other values merely by suitably altering the input voltage u).

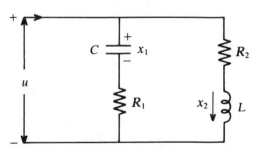

Fig. 4.1 Electric circuit for Problem 4.14.

57

Some further applications of determinants are included in the exercises at the end of this chapter.

4.2 Evaluation of determinants

As remarked in Section 4.1.2, the definition (4.23) does not provide a feasible way of evaluating large determinants with numerical elements. The reason is easily seen: the expansion of a 3×3 determinant involves three 2×2 determinants, so involves $3 \times 2 = 3!$ products; similarly a 4×4 determinant involves $4 \times 3 \times 2 = 4!$ products, etc., and generally an $n \times n$ determinant contains $n!$ products if expanded by repeated use of (4.23) or (4.24). The number $n!$ increases very rapidly as n increases, for example $18! \approx 6.4 \times 10^{15}$. Assuming a computer multiplication time of 10^{-6} sec, an 18×18 determinant evaluated in this way would take approximately 200 years! Clearly an alternative method must be used.

The easiest way to proceed is to reduce a given $n \times n$ matrix A to triangular form using gaussian elimination, as described in Section 3.2. The operations which are performed on the *rows* of A to achieve the triangularization can now be recognized as precisely those described in properties PD6(a) and (c). Operations of type (c) leave $\det A$ unaltered, but those of type (a) introduce a factor -1. The final triangular determinant is equal, by PD5, to the product of the elements on the principal diagonal (i.e., the product of the pivots) so $\det A$ is equal to this product multiplied by $(-1)^k$, where k is the number of row interchanges (these are introduced when partial pivoting is used). It can be shown that the number of multiplications involved in gaussian elimination is of the order of $\frac{1}{3}n^3$ for an $n \times n$ determinant, so computer evaluation when $n = 18$ would take only about 2×10^{-3} sec.

Example 4.11 Let A be the matrix in Eq. (4.19). To reduce the first column of $\det A$ we apply $(R2) - 2(R1)$, $(R3) - (-3)(R1)$ so that (see (4.31))

$$\det A = \begin{vmatrix} 1 & -3 & 7 \\ 0 & 10 & -17 \\ 0 & -2 & 23 \end{vmatrix}$$

$$= \begin{vmatrix} 1 & -3 & 7 \\ 0 & 10 & -17 \\ 0 & 0 & \dfrac{196}{10} \end{vmatrix}, \qquad (R3) - (-2/10)(R2)$$

$$= 1 \times 10 \times \frac{196}{10} = 196$$

which agrees with the value found in Example 4.1.

In fact the matrix A in (4.19) is the matrix of Eqs (3.10). The elimination operations used to reduce $\det A$ above are exactly the same as those used to reduce the augmented matrix B in (3.15) and (3.16).

It can now be appreciated that when gaussian elimination is performed on a system of equations $Ax = b$, the determinant of A is obtained as a by-product of the calculations, provided the number of row interchanges is recorded.

Problem 4.15 From your working for Problems 3.2 and 3.3 write down the values of the determinants of the corresponding matrices A.

Problem 4.16 Using gaussian elimination evaluate

$$\text{(a)} \quad \begin{vmatrix} 1 & 1 & 1 \\ 35 & 37 & 34 \\ 23 & 26 & 25 \end{vmatrix} \qquad \text{(b)} \quad \begin{vmatrix} 1 & 2 & 3 & 4 \\ -1 & 1 & 2 & 3 \\ 1 & -1 & 1 & 2 \\ -1 & 1 & -1 & 5 \end{vmatrix}$$

It will be recalled that the gaussian elimination method breaks down if at any stage it is impossible to find a nonzero pivot. In this case $\det A = 0$, as the following argument shows.

Example 4.12 Suppose that at the third stage of reducing a 4×4 determinant we obtain

$$\begin{vmatrix} k_1 & k_2 & k_4 & k_6 \\ 0 & k_3 & k_5 & k_7 \\ 0 & 0 & 0 & k_8 \\ 0 & 0 & 0 & k_9 \end{vmatrix} \qquad (4.38)$$

It is impossible to obtain a nonzero pivot in the 3,3 position by row interchanges. Expanding (4.38) by the first column twice in succession gives

$$k_1 k_3 \begin{vmatrix} 0 & k_8 \\ 0 & k_9 \end{vmatrix}$$

which is zero, because of the zero column (see Problem 4.6).

A square matrix A for which $\det A = 0$ is called *singular*, otherwise if $\det A \neq 0$ then A is *nonsingular*.

59

Problem 4.17 Prove using gaussian elimination that

$$\det \begin{bmatrix} 1 & -2 & 1 & -1 \\ 1 & 5 & -7 & 2 \\ 3 & 1 & -5 & 3 \\ 2 & 3 & -6 & 0 \end{bmatrix} = 0$$

It should be noted that the matrix in Problem 4.17 is that of the equations in Problem 3.4 which were found to be inconsistent. This again illustrates the point that the equations (3.2), i.e., $Ax = b$, have a unique solution only if $\det A \neq 0$ (a formal treatment of this important result will be given in Section 5.4.3).

Problem 4.18 If X and Y are two arbitrary $n \times n$ nonsingular matrices prove that each of $X^T X$, XX^T, XY, YX, and X^k ($k = 2, 3, 4, \ldots$) is also nonsingular.

Problem 4.19 Let A be a general $n \times n$ 'reversed lower triangular' matrix, having all elements above the secondary diagonal equal to zero, e.g., when $n = 3$

$$A = \begin{bmatrix} 0 & 0 & a_{13} \\ 0 & a_{22} & a_{23} \\ a_{31} & a_{32} & a_{33} \end{bmatrix}$$

Evaluate $\det A$ by transforming it into usual lower triangular form. In particular, what is $\det J_n$, where J_n is the matrix defined in Exercise 2.4?

4.3 Inverse

4.3.1 Definition and properties

Following Eq. (4.8), the *inverse B* of an $n \times n$ matrix A is defined to be a matrix satisfying

$$AB = BA = I_n \tag{4.39}$$

Suppose there exist *two* matrices B and B_1 which both satisfy (4.39). Then $AB_1 = I$, which on premultiplication by B gives $BAB_1 = B$, so $IB_1 = B$ in view of (4.39), whence $B_1 = B$. Hence, if an inverse does exist then it is *unique*. We have already encountered the notation A^{-1} for the inverse of A. Next, application of (4.33) to (4.39) gives

$$\det(AB) = (\det A)(\det B) = \det I_n = 1 \tag{4.40}$$

the latter part of (4.40) following by use of (4.29). Now since B is an

$n \times n$ matrix with finite elements, $\det B$ must also be a finite number, so if $\det A = 0$ then (4.40) implies that B does not exist. In other words, A^{-1} exists only if A is nonsingular, in which case (4.39) can be written

$$AA^{-1} = A^{-1}A = I_n \qquad (4.41)$$

and (4.40) implies

$$\det(A^{-1}) = \frac{1}{\det A} \qquad (4.42)$$

The converse also applies, namely that if A is nonsingular then A^{-1} exists. In fact this can be demonstrated by giving an explicit formula for A^{-1}. For simplicity we detail the case $n = 3$. Consider the product AC where

$$C = \begin{bmatrix} A_{11} & A_{21} & A_{31} \\ A_{12} & A_{22} & A_{32} \\ A_{13} & A_{23} & A_{33} \end{bmatrix} \qquad (4.43)$$

and the A_{ij} are the cofactors defined in Section 4.1. Using (4.23) and (4.25) it is easily verified that

$$AC = \begin{bmatrix} |A| & 0 & 0 \\ 0 & |A| & 0 \\ 0 & 0 & |A| \end{bmatrix} \qquad (4.44)$$

For example, the 1,1 element of AC is the term in (4.18); the 1,3 element of AC is the term in (4.21). Thus

$$AC = (|A|)I_3 \qquad (4.45)$$

and since $|A|$ is assumed nonzero we can divide in (4.45) to obtain

$$A\left(\frac{C}{|A|}\right) = I_3$$

Similarly, it can be verified that $(C/|A|)A = I_3$, using (4.24) and (4.26). Thus the matrix $C/|A|$ is the inverse of A since it satisfies (4.39). The matrix C is called the *adjoint* of A, usually denoted by adj$A$. It is the *transpose* of the matrix of cofactors (note that in (4.43) the cofactors of the rows of A form the columns of C). The argument generalizes immediately for any value of n, so we have shown that if A is nonsingular its inverse is

$$A^{-1} = \left(\frac{1}{\det A}\right)\text{adj}A \qquad (4.46)$$

Example 4.13 We use (4.46) to calculate the inverse of

$$A = \begin{bmatrix} 2 & -1 & 1 \\ 1 & 0 & 1 \\ 3 & -1 & 4 \end{bmatrix} \qquad (4.47)$$

Using the definition (4.17) the cofactors of the first row are

$$A_{11} = + \begin{vmatrix} 0 & 1 \\ -1 & 4 \end{vmatrix} = 1, \qquad A_{12} = - \begin{vmatrix} 1 & 1 \\ 3 & 4 \end{vmatrix} = -1, \qquad A_{13} = + \begin{vmatrix} 1 & 0 \\ 3 & -1 \end{vmatrix} = -1$$

which gives the first column of adj$A$, and expanding by the first row

$$\det A = 2 \cdot 1 + (-1)(-1) + 1 \cdot (-1) = 2$$

The other cofactors are evaluated similarly, and substitution into (4.46) gives

$$A^{-1} = \frac{1}{2} \begin{bmatrix} 1 & 3 & -1 \\ -1 & 5 & -1 \\ -1 & -1 & 1 \end{bmatrix} \tag{4.48}$$

In general (4.46) is useless as a numerical method for finding A^{-1} since it involves calculation of n^2 determinants, each having order $n-1$. We shall discuss practical evaluation of A^{-1} in Section 4.4. The special case of (4.46) when $n = 2$ was given in Eq. (4.7) and is worth remembering.

Problem 4.20 Verify by direct multiplication that if D is the diagonal matrix in (2.36) then

$$D^{-1} = \text{diag}[1/d_{11}, 1/d_{22}, \ldots, 1/d_{nn}] \tag{4.49}$$

provided all $d_{ii} \neq 0$.

Problem 4.21 Use (4.46) to calculate the inverse of

$$A = \begin{bmatrix} 2 & 1 & -5 \\ 3 & 2 & 4 \\ 1 & 0 & 3 \end{bmatrix} \tag{4.50}$$

Going back to the discussion at the beginning of Chapter 3, we can now settle the question of 'division': provided $|A| \neq 0$ the unique solution of the equations $Ax = b$ is given by $x = A^{-1}b$.

Example 4.14 If A is the matrix in (4.47) and $b = [2, 4, -1]^T$, the solution of $Ax = b$ is

$$x = A^{-1} \begin{bmatrix} 2 \\ 4 \\ -1 \end{bmatrix} = \frac{1}{2} \begin{bmatrix} 15 \\ 19 \\ -7 \end{bmatrix}$$

using the expression for A^{-1} in (4.48).

DETERMINANT AND INVERSE

Problem 4.22 Using the result of Problem 4.21 solve the equations

$$2x_1 + x_2 - 5x_3 = 1$$
$$3x_1 + 2x_2 + 4x_3 = 0$$
$$x_1 \qquad + 3x_3 = 2$$

We can also now analyse a difficulty encountered in Section 2.2.3. We saw that if $AC = AD$, it does not necessarily follow that $C = D$. However, if A is nonsingular we have

$$A^{-1}AC = A^{-1}AD$$

which gives

$$IC = ID$$

whence $C = D$. Thus the concept of inverse allows us to manipulate equations involving matrices in a manner analogous to ordinary algebra.

As a further example, which is of considerable importance, let A and B be two $n \times n$ nonsingular matrices and let $X = (AB)^{-1}$. This implies from (4.39) that $XAB = I$, so we can construct the following sequence of identities:

$$XABB^{-1} = IB^{-1} = B^{-1}$$
$$XAI = B^{-1}, \qquad XAA^{-1} = B^{-1}A^{-1}, \qquad XI = B^{-1}A^{-1}$$

showing that $X = B^{-1}A^{-1}$, i.e.

$$(AB)^{-1} = B^{-1}A^{-1} \tag{4.51}$$

Note again the reversal of order, as encountered previously in the result for $(AB)^T$ in (2.41).

We can similarly find the inverse of the transpose of a matrix. Writing $X = (A^T)^{-1}$ we have

$$A^T X = I \tag{4.52}$$

and taking the transpose of both sides of (4.52) gives, using (2.38) and (2.41)

$$X^T A = I^T = I \tag{4.53}$$

Equation (4.53) implies that $X^T = A^{-1}$, and transposing this gives $(X^T)^T = X = (A^{-1})^T$, so we have proved

$$(A^T)^{-1} = (A^{-1})^T \tag{4.54}$$

The identity (4.54) states that the order of the operations of transposing and inverting a matrix can be interchanged.

Problem 4.23 If A, B, C are three $n \times n$ nonsingular matrices, prove that $(ABC)^{-1} = C^{-1}B^{-1}A^{-1}$.

63

Problem 4.24 Prove that $(A^2)^{-1} = (A^{-1})^2$, and similarly show that $(A^k)^{-1} = (A^{-1})^k$ for any positive integer k (thus we can write $(A^k)^{-1}$ as A^{-k}).

Problem 4.25 If B is nonsingular and $AB^{-1} = B^{-1}A$, what condition must A and B satisfy?

Problem 4.26 Use (2.75) to prove that if A and B are two $n \times n$ nonsingular matrices then $(A \otimes B)^{-1} = A^{-1} \otimes B^{-1}$.
Notice that in comparison with (4.51) the Kronecker product formula does not involve reversal of order (as for the transpose formula, (2.68)).

Problem 4.27 Prove that if A is nonsingular and symmetric then A^{-1} is also symmetric.

Problem 4.28 If A is a nonsingular matrix with complex elements, prove that $(A*)^{-1} = (A^{-1})*$.

Problem 4.29 If A is the matrix in (4.50), find the matrix X such that $A^T X A = I$.

4.3.2 Partitioned form

It is sometimes useful to obtain the inverse of a matrix in partitioned form. For example, suppose

$$F = \begin{bmatrix} A & B \\ C & D \end{bmatrix} \begin{matrix} n \\ k \end{matrix} \qquad \begin{matrix} n & k \end{matrix}$$

and let

$$F^{-1} = \begin{bmatrix} W & X \\ Y & Z \end{bmatrix} \begin{matrix} n \\ k \end{matrix} \qquad \begin{matrix} n & k \end{matrix}$$

By definition $FF^{-1} = I_{n+k}$ so

$$\begin{bmatrix} A & B \\ C & D \end{bmatrix}\begin{bmatrix} W & X \\ Y & Z \end{bmatrix} = \begin{bmatrix} I_n & 0 \\ 0 & I_k \end{bmatrix} \tag{4.55}$$

and applying the rules of partitioned multiplication (Section 2.4) to (4.55) produces

$$AW + BY = I_n \tag{4.56}$$

$$AX + BZ = 0 \tag{4.57}$$

$$CW + DY = 0 \tag{4.58}$$

$$CX + DZ = I_k \tag{4.59}$$

Suppose that F has been partitioned so that A is nonsingular. Premultiplication of (4.56) by A^{-1} gives

$$W = A^{-1} - A^{-1}BY \qquad (4.60)$$

and substitution into (4.58) gives

$$CA^{-1} - CA^{-1}BY + DY = 0$$

from which

$$Y = -(D - CA^{-1}B)^{-1}CA^{-1} \qquad (4.61)$$

provided $G = D - CA^{-1}B$ is nonsingular.

Similarly from (4.57) and (4.59)

$$Z = G^{-1}, \qquad X = -A^{-1}BG^{-1} \qquad (4.62)$$

An advantage of using (4.60), (4.61) and (4.62) to calculate the inverse of the $(n + k) \times (n + k)$ matrix F is that the matrices to be inverted (A and G) have orders only n and k.

Problem 4.30 Obtain the inverse of A in (4.47) using the partitioned form

$$\begin{bmatrix} 2 & -1 & 1 \\ 1 & 0 & 1 \\ \hline 3 & -1 & 4 \end{bmatrix}$$

4.4 Calculation of inverse

As stated at the beginning of Chapter 3, A^{-1} is best evaluated by computing the solution of associated sets of linear equations. If $X = A^{-1}$ and the columns of X are denoted by $X_1, X_2, \ldots, X_n$ then $AX = I$ implies

$$AX_i = e_i^T, \qquad i = 1, 2, \ldots, n \qquad (4.63)$$

where e_i^T denotes the ith column of I_n (see Problem 2.19). The equations (4.63) show that the ith column X_i of A^{-1} is the solution of the set of Eqs (3.2) with right-hand side equal to the ith column of I_n. This solution can be determined by gaussian elimination and back substitution.

Example 4.15 Let us determine the second column of the inverse of A in (4.19), which is the matrix of Eqs (3.10). Replace the right-hand sides of Eqs (3.10) by 0, 1, 0 so the augmented matrix (3.14) is $B = [A, e_2^T]$, i.e.,

$$B = \begin{bmatrix} 1 & -3 & 7 & 0 \\ 2 & 4 & -3 & 1 \\ -3 & 7 & 2 & 0 \end{bmatrix} \rightarrow \begin{bmatrix} 1 & -3 & 7 & 0 \\ 0 & 10 & -17 & 1 \\ 0 & 0 & \dfrac{196}{10} & \dfrac{2}{10} \end{bmatrix} \qquad (4.64)$$

using the same operations as in (3.15) and (3.16). The equations corresponding to (4.64) are

$$\begin{aligned} x_1 - 3x_2 + 7x_3 &= 0 \\ 10x_2 - 17x_3 &= 1 \\ 196x_3 &= 2 \end{aligned}$$
(4.65)

and back substitution gives $x_3 = 2/196$, $x_2 = 23/196$, $x_1 = 55/196$, so the second column of A^{-1} is $(1/196)[55, 23, 2]^T$.

The transformations on each of the columns e_i^T can be carried out simultaneously by using an augmented matrix

$$B = [A \mid I_n]$$
(4.66)

However, each column of A^{-1} must be found separately by back substitution on a set of equations like (4.65). The following modified form of gaussian elimination (called *Gauss–Jordan elimination*) is more convenient when carrying out calculations by hand.

The basic idea is to take the reduced triangular form of the augmented matrix and apply *further* row operations until the array of the first n columns is *diagonal*, with unit pivots. No back substitution is then required. The procedure is first illustrated for B in (4.64).

Example 4.15 (continued) Starting with the triangular form in (4.64), first reduce the pivots to unity by

$$\left(\frac{1}{10}\right)(R2), \qquad \left(\frac{10}{196}\right)(R3)$$
(4.67)

giving

$$\begin{bmatrix} 1 & -3 & 7 & \vdots & 0 \\ 0 & 1 & -\dfrac{17}{10} & \vdots & \dfrac{1}{10} \\ 0 & 0 & 1 & \vdots & \dfrac{2}{196} \end{bmatrix}$$
(4.68)

Then reduce the upper part of the first three columns of (4.68), working from left to right:

$$\rightarrow \begin{bmatrix} 1 & 0 & \dfrac{19}{10} & \vdots & \dfrac{3}{10} \\ 0 & 1 & -\dfrac{17}{10} & \vdots & \dfrac{1}{10} \\ 0 & 0 & 1 & \vdots & \dfrac{2}{196} \end{bmatrix} \quad (R1) + 3(R2)$$
(4.69)

$$\rightarrow \begin{bmatrix} 1 & 0 & 0 & \frac{55}{196} \\ 0 & 1 & 0 & \frac{23}{196} \\ 0 & 0 & 1 & \frac{2}{196} \end{bmatrix} \qquad \begin{array}{l} (R1) - \left(\frac{19}{10}\right)(R3) \\ \\ (R2) + \left(\frac{17}{10}\right)(R3) \end{array} \qquad (4.70)$$

This is equivalent to converting the equations (4.65) into

$$x_1 \qquad\quad = \frac{55}{196}$$
$$\quad x_2 \qquad = \frac{23}{196}$$
$$\qquad x_3 \;\; = \frac{2}{196}$$

so the last column in (4.70) is precisely the desired second column of A^{-1}. *All* the columns of A^{-1} can be obtained simultaneously by commencing with the augmented matrix (4.66). Continuing with Example 4.15:

$$B = \begin{bmatrix} 1 & -3 & 7 & 1 & 0 & 0 \\ 2 & 4 & -3 & 0 & 1 & 0 \\ -3 & 7 & 2 & 0 & 0 & 1 \end{bmatrix}$$

$$\rightarrow \begin{bmatrix} 1 & -3 & 7 & 1 & 0 & 0 \\ 0 & 10 & -17 & -2 & 1 & 0 \\ 0 & 0 & \frac{196}{10} & \frac{26}{10} & \frac{2}{10} & 1 \end{bmatrix} \qquad \begin{array}{l} (R2) - 2(R1) \\ (R3) + 3(R1) \\ (R3) + \frac{2}{10}(R2) \end{array} \qquad (4.71)$$

these steps being just the gaussian elimination used to obtain (4.64). The operations in (4.67), (4.69), and (4.70) are then applied to (4.71), giving the following (only the effect on the last three columns is presented, the reduction of the first three columns is as before, in (4.68), (4.69), and (4.70)).

$$\begin{array}{c} \frac{1}{10}(R2) \\ \xrightarrow{\hspace{1cm}} \\ \frac{10}{196}(R3) \end{array} \begin{bmatrix} 1 & 0 & 0 \\ \frac{-2}{10} & \frac{1}{10} & 0 \\ \frac{26}{196} & \frac{2}{196} & \frac{10}{196} \end{bmatrix} \quad \xrightarrow{(R1)+3(R2)} \begin{bmatrix} \frac{4}{10} & \frac{3}{10} & 0 \\ \frac{-2}{10} & \frac{1}{10} & 0 \\ \frac{26}{196} & \frac{2}{196} & \frac{10}{196} \end{bmatrix}$$

$$\begin{array}{c} (R1) - \left(\frac{19}{10}\right)(R3) \\ \xrightarrow{\hspace{2cm}} \\ (R2) + \left(\frac{17}{10}\right)(R3) \end{array} \begin{bmatrix} \frac{29}{196} & \frac{55}{196} & \frac{-19}{196} \\ \frac{5}{196} & \frac{23}{196} & \frac{17}{196} \\ \frac{26}{196} & \frac{2}{196} & \frac{10}{196} \end{bmatrix}$$

Thus the inverse of A in (4.19) is

$$A^{-1} = \frac{1}{196} \begin{bmatrix} 29 & 55 & -19 \\ 5 & 23 & 17 \\ 26 & 2 & 10 \end{bmatrix} \qquad (4.72)$$

Summarizing, to compute the inverse of an $n \times n$ nonsingular matrix A:
(a) form the $n \times 2n$ matrix B in (4.66);
(b) reduce the first n columns of B to upper triangular form by gaussian elimination;
(c) divide each row of the reduced matrix by the pivot for that row;
(d) reduce the upper part of the first n columns to zero, starting with column 2 and working from left to right;
(e) the last n columns give A^{-1}.

Symbolically we can write

$$[A \mid I_n] \to [I_n \mid A^{-1}]$$

Notice that *any* set of equations $Ax = b$ could be solved by Gauss–Jordan elimination instead of ordinary gaussian elimination. However, the latter method is generally to be preferred in this case, since it involves a smaller total computational effort.

Problem 4.31 Calculate the inverse of the matrices in Problem 3.2 and in (4.50) using the Gauss–Jordan method.

Problem 4.32 Calculate the inverse of the 4×4 coefficient matrix in Problem 3.6 using the Gauss–Jordan procedure.

Notice that the Gauss–Jordan method is particularly simple when applied to an $n \times n$ triangular matrix (upper or lower), since only half the procedure is needed. For example, consider a 3×3 upper triangular nonsingular matrix U. Applying the preceding rules (c) and (d) we obtain

$$[U \mid I] \to \begin{bmatrix} 1 & \dfrac{u_2}{u_1} & \dfrac{u_3}{u_1} & \dfrac{1}{u_1} & 0 & 0 \\[2ex] 0 & 1 & \dfrac{u_5}{u_4} & 0 & \dfrac{1}{u_4} & 0 \\[2ex] 0 & 0 & 1 & 0 & 0 & \dfrac{1}{u_6} \end{bmatrix} \begin{matrix} \dfrac{1}{u_1}(R1) \\[2ex] \dfrac{1}{u_4}(R2) \\[2ex] \dfrac{1}{u_6}(R3) \end{matrix}$$

$$\to \begin{bmatrix} 1 & 0 & \alpha & \dfrac{1}{u_1} & -\dfrac{u_2}{u_1 u_4} & 0 \\[2ex] 0 & 1 & \dfrac{u_5}{u_4} & 0 & \dfrac{1}{u_4} & 0 \\[2ex] 0 & 0 & 1 & 0 & 0 & \dfrac{1}{u_6} \end{bmatrix} \quad (R1) - \dfrac{u_2}{u_1}(R2)$$

68

where $\alpha = (u_3/u_1) - (u_2/u_1)(u_5/u_4)$, and the operations $(R1) - \alpha(R3)$ and $(R2) - (u_5/u_4)(R3)$ then give U^{-1}. From the nature of the operations it is obvious that U^{-1} will also be upper triangular, and this applies for arbitrary n. Similarly, the inverse of an arbitrary nonsingular lower triangular matrix is lower triangular. In either case it is not difficult to derive algebraic formulae for the elements of the inverse for arbitrary n.

Problem 4.33 Complete the above derivation of U^{-1}. Similarly, obtain the inverse of the 3×3 lower triangular matrix in (3.22).

It is interesting to link the Gauss–Jordan method with the triangular decomposition $A = LU$ of Section 3.3. The equation $AX = I$ is equivalent to $LUX = I$, and the first (gaussian) part of the process produces $UX = L^{-1}$, so the reduced array at this stage is $[U \mid L^{-1}]$. For example, for the matrix A in (4.19) used in Example 4.15, the first three columns in (4.71) are identical to the matrix U in Eq. (3.29), and the last three columns in (4.71) give the inverse L^{-1} of the matrix L in Eq. (3.29). Since $A^{-1} = (LU)^{-1} = U^{-1}L^{-1}$, on inverting U (as described above) we can therefore express A^{-1} as a product of triangular factors.

Problem 4.34 Calculate U^{-1} for the matrix A in Example 4.15, and hence express A^{-1} as a product of triangular factors. Check, by evaluating the product, that your expression agrees with (4.72).

Repeat this process for A in (4.50).

4.5 Cramer's rule

Consider again the n equations in n unknowns, $Ax = b$, when they have the unique solution

$$x = A^{-1}b \qquad (4.73)$$

Substitute the expression (4.46) for A^{-1} into (4.73), giving

$$x = \left(\frac{1}{\det A}\right)(\text{adj}A)b \qquad (4.74)$$

For simplicity take $n = 3$, and use the expression for $\text{adj}A$ given in (4.43). Then (4.74) gives

$$x_1 = \left(\frac{1}{\det A}\right)(A_{11}b_1 + A_{21}b_2 + A_{31}b_3) \qquad (4.75)$$

with similar expressions for x_2 and x_3. The interesting feature to notice about (4.75) is that the numerator is precisely the determinant

$$\begin{vmatrix} b_1 & a_{12} & a_{13} \\ b_2 & a_{22} & a_{23} \\ b_3 & a_{32} & a_{33} \end{vmatrix} \qquad (4.76)$$

expanded by the first column, since the cofactors of this first column are A_{11}, A_{21}, A_{31}, the same as the cofactors of the first column of $\det A$. The generalization of (4.75) and (4.76) is similarly obtained as

$$x_i = \left(\frac{1}{\det A}\right) \times \left(\begin{array}{l}\text{determinant of the matrix obtained} \\ \text{from } A \text{ by replacing the } i\text{th column by } b\end{array}\right) \qquad (4.77)$$

for $i = 1, 2, \ldots, n$. This result is known as *Cramer's rule*. It is occasionally useful for small values of n when not all of the coefficients in the equations are known numerically, but is worthless as a numerical method since it involves calculating $n + 1$ determinants, each having order n.

Example 4.16 For Eqs (3.10) in Example 3.4, Cramer's rule gives

$$x_1 = \frac{1}{196} \begin{vmatrix} 2 & -3 & 7 \\ -1 & 4 & -3 \\ 3 & 7 & 2 \end{vmatrix} = -\frac{27}{98}$$

$$x_2 = \frac{1}{196} \begin{vmatrix} 1 & 2 & 7 \\ 2 & -1 & -3 \\ -3 & 3 & 2 \end{vmatrix} = \frac{19}{98}$$

$$x_3 = \frac{1}{196} \begin{vmatrix} 1 & -3 & 2 \\ 2 & 4 & -1 \\ -3 & 7 & 3 \end{vmatrix} = \frac{20}{49}$$

($\det A$ was calculated to be 196 in Example 4.1.)

Problem 4.35 Solve the equations in Problem 3.2 by Cramer's rule ($\det A$ has been evaluated in Problem 4.15).

Exercises

4.1 It can be shown that the quadratic equations $a_0\lambda^2 + a_1\lambda + a_2 = 0$ and $b_0\lambda^2 + b_1\lambda + b_2 = 0$ ($a_0 \neq 0$, $b_0 \neq 0$) have a common root if and only if

$$\begin{vmatrix} a_0 & a_1 & a_2 & 0 \\ 0 & a_0 & a_1 & a_2 \\ 0 & b_0 & b_1 & b_2 \\ b_0 & b_1 & b_2 & 0 \end{vmatrix} = 0$$

(this determinant is a special case of the *resultant* of two polynomials, and is important in the theory of equations). Use this result to show that the

70

equations

$$ax^2 + x + (1-a) = 0, \qquad (1-b)x^2 + x + b = 0$$

have a common root for *any* values of a and b.

4.2 An equation with real coefficients

$$a_0\lambda^3 + a_1\lambda^2 + a_2\lambda + a_3 = 0$$

($a_0 \neq 0$) has all its roots with negative real parts if and only if the following conditions are satisfied:

$$a_1 > 0, \qquad \begin{vmatrix} a_1 & a_3 \\ a_0 & a_2 \end{vmatrix} > 0, \qquad \begin{vmatrix} a_1 & a_3 & 0 \\ a_0 & a_2 & 0 \\ 0 & a_1 & a_3 \end{vmatrix} > 0$$

These conditions are important in the study of stability of linear continuous-time systems (like that in Example 1.7), and can be generalized for nth-degree equations.

Determine for what values of k the equation

$$(3-k)\lambda^3 + 2\lambda^2 + (5-2k)\lambda + 2 = 0$$

has all its roots with negative real parts.

4.3 The equations

$$i_1 R_6 + (i_1 - i_2)R_3 + (i_1 - i_3)R_4 = e$$
$$(i_1 - i_3)R_4 - i_3 R_1 - (i_3 - i_2)R_5 = 0$$
$$(i_1 - i_2)R_3 + (i_3 - i_2)R_5 - i_2 R_2 = 0$$

describe the Wheatstone bridge shown in Fig. 4.2. Obtain expressions for i_2 and i_3 as ratios of determinants using Cramer's rule. Hence show that if there is no current through the galvanometer (i.e., $i_2 = i_3$) then $R_1 R_3 = R_2 R_4$.

4.4 In Example 1.7 let a force $u(t)$ be applied to the right-hand mass as shown in Fig. 1.4. This leads to a term $[0, 0, 1/m_1, 0]^T u$ added to the right-hand side of Eqs (1.10). If $m_1 = m_2 = 1$, $d_1 = d_2 = 1$, and $k_2 = \frac{1}{4}$, use Eq. (4.37) to show that the condition for the system to be controllable is $k_1 \neq \frac{1}{2}$.

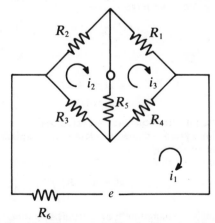

Fig. 4.2 **Wheatstone bridge for Exercise 4.3.**

71

4.5 If

$$M = \begin{matrix} n & m \\ \begin{bmatrix} A & 0 \\ C & D \end{bmatrix} & \begin{matrix} n \\ m \end{matrix} \end{matrix} \tag{4.78}$$

then M is called *lower block triangular.*

(a) If A is a lower triangular matrix, prove that

$$\det M = (a_{11}a_{22} \cdots a_{nn}) \det D$$

(b) Hence deduce that if A is an *arbitrary* $n \times n$ matrix then $\det M = \det A \det D$.

4.6 If in (4.78) A and D are both nonsingular, obtain M^{-1} in the same partitioned form.

4.7 Let

$$U = \begin{matrix} n & m \\ \begin{bmatrix} W & X \\ Y & Z \end{bmatrix} & \begin{matrix} n \\ m \end{matrix} \end{matrix}, \qquad V = \begin{bmatrix} W^{-1} & 0 \\ -YW^{-1} & I_m \end{bmatrix}$$

where W is nonsingular. By applying PD7 to the product VU and using the result of Exercise 4.5(b), prove that

$$\det U = \det W \det(Z - YW^{-1}X) \tag{4.79}$$

Similarly, it can be shown that if Z is nonsingular then $\det U = \det Z \det(W - XZ^{-1}Y)$.

4.8 Consider Eqs (4.36) describing a linear control system. In practice it may be possible to measure only a scalar *output*

$$y = cx = c_1x_1 + c_2x_2 + \cdots + c_nx_n$$

If $\bar{x}(s) = \mathcal{L}\{x(t)\}$ denotes the Laplace transform of $x(t)$, then the transform of dx/dt is $s\bar{x}$, assuming $x(0) = 0$ (see, e.g., Jones (1976), Chapter 7).

Taking the Laplace transform of (4.36) thus gives

$$s\bar{x} = A\bar{x} + b\bar{u}$$

and on rearrangement we have

$$(sI_n - A)\bar{x} = b\bar{u}$$

Hence

$$\bar{y} = c\bar{x}$$
$$= c(sI_n - A)^{-1}b\bar{u}$$
$$= g(s)\bar{u}$$

where the rational function $g(s) = c(sI_n - A)^{-1}b$ is called the *transfer function* of the system, since it relates the Laplace transform of the output y to that of the input u.

Use (4.79) to prove that

$$g(s) = \begin{vmatrix} (sI - A) & \overset{b}{B} \\ -c & 0 \end{vmatrix} \Big/ |sI - A|$$

4.9 Prove that if A is nonsingular then $\det(\mathrm{adj}A) = (\det A)^{n-1}$ (hint: use PD2 and PD7). Show also that $\mathrm{adj}(A^{T}) = (\mathrm{adj}A)^{T}$.

72

4.10 A real $n \times n$ matrix U is called *orthogonal* if $U^T U = I_n$. Prove
(a) The 2×2 matrix in Eq. (1.7) is orthogonal.
(b) $U^{-1} = U^T$.
(c) $\det U = \pm 1$.
(d) U^T and U^{-1} are orthogonal.
(e) The product of any two $n \times n$ orthogonal matrices is also orthogonal.
(f) $\dfrac{1}{\sqrt{2}}\begin{bmatrix} U & U \\ -U & U \end{bmatrix}$ is a $2n \times 2n$ orthogonal matrix.

4.11 *A complex $n \times n$ matrix U is called unitary if $U^*U = I_n$.*
(a) Prove that $|\det U| = 1$.
(b) If S is an arbitrary $n \times n$ skew hermitian matrix it can be shown (see Exercise 6.2) that $I_n + S$ is nonsingular. Prove that $(I_n - S)(I_n + S)^{-1}$ is unitary.
(c) If U is such that $I_n + U$ is nonsingular, prove that $(I_n - U)(I_n + U)^{-1}$ is skew hermitian.

4.12 If A and B are two $n \times n$ orthogonal matrices, prove that $A + B = A(A^T + B^T)B$. Hence prove that if $\det A + \det B = 0$, then $A + B$ is singular.

4.13 If P is an $n \times n$ nonsingular matrix and $B = P^{-1}AP$, prove that $B^k = P^{-1}A^k P$ for any positive integer k. If

$$A = \begin{bmatrix} 2 & 1 \\ 1 & 2 \end{bmatrix}, \qquad P = \begin{bmatrix} 1 & 1 \\ -1 & 1 \end{bmatrix}$$

verify that B is diagonal. Write down B^5 and hence calculate A^5.

4.14 If X is an $n \times n$ matrix satisfying the equation $X^2 = X$, it is called *idempotent*. Deduce that either $X = I_n$ or X is singular.

4.15 If A is an $n \times n$ matrix such that $A^k = 0$, $A^{k-1} \neq 0$ for some positive integer k, it is called *nilpotent* (an example was given in Exercise 2.8). Prove (a) A is singular, (b) $(I - A)^{-1} = I + A + A^2 + \cdots + A^{k-1}$.

4.16 The $n \times n$ *Vandermonde* matrix V_n is defined by

$$V_n = \begin{bmatrix} 1 & 1 & \dots 1 \\ \lambda_1 & \lambda_2 & \dots \lambda_n \\ \lambda_1^2 & \lambda_2^2 & \dots \lambda_n^2 \\ \vdots & \vdots & \vdots \\ \lambda_1^{n-1} & \lambda_2^{n-1} & \dots \lambda_n^{n-1} \end{bmatrix} \tag{4.80}$$

Show that $\det V_2 = (\lambda_2 - \lambda_1)$, and by reducing the last column of $\det V_3$ to $[1, 0, 0]^T$, that

$$\det V_3 = (\lambda_3 - \lambda_2)(\lambda_3 - \lambda_1) \det V_2,$$

Similarly, prove that

$$\det V_n = (\lambda_n - \lambda_{n-1})(\lambda_n - \lambda_{n-2}) \cdots (\lambda_n - \lambda_1) \det V_{n-1}$$

and hence deduce that

$$\det V_n = \prod_{n \geqslant j > i \geqslant 1} (\lambda_j - \lambda_i) \tag{4.81}$$

Thus the Vandermonde matrix is nonsingular if and only if all the λ's are different from each other.

73

4.17 It is required to determine a polynomial

$$y = a_0 + a_1 x + a_2 x^2 + \cdots + a_{n-1} x^{n-1}$$

so that it passes through n given points $(x_1, y_1), (x_2, y_2), \ldots, (x_n, y_n)$. Show that the coefficients are determined by $aV = y$, where $a = [a_0, a_1, \ldots, a_{n-1}]$, $y = [y_1, \ldots, y_n]$ and V is a Vandermonde matrix in the form (4.80). (Notice that the solution for a is unique if all the x's are different, because of (4.81)).

4.18 If $A = \text{diag}[\lambda_1, \lambda_2, \ldots, \lambda_n]$ and $b = [b_1, b_2, \ldots, b_n]^T$ show that the controllability matrix in (4.37) is equal to

$$(\text{diag}[b_1, b_2, \ldots, b_n]) V_n^T$$

where V_n is defined in (4.80). Hence deduce that in this case Eq. (4.36) is controllable if and only if all the λ's are different from each other and all $b_i \neq 0$.

4.19 If A is nonsingular and dA/dt and $d(A^{-1})/dt$ both exist, use Eq. (2.70) to prove that

$$\frac{d(A^{-1})}{dt} = -A^{-1} \frac{dA}{dt} A^{-1}$$

(this corresponds to $d(a^{-1})/dt = -a^{-2} \, da/dt$ for a scalar function $a(t)$).

4.20 If the elements of a 2×2 matrix A are differentiable functions of t, prove that

$$\frac{d}{dt}(\det A) = \begin{vmatrix} \dot{a}_{11} & \dot{a}_{12} \\ a_{21} & a_{22} \end{vmatrix} + \begin{vmatrix} a_{11} & a_{12} \\ \dot{a}_{21} & \dot{a}_{22} \end{vmatrix}$$

This can be extended to an $n \times n$ matrix:

$$\frac{d}{dt}(\det A) = \sum_{i=1}^{n} \det A_i$$

where A_i is the matrix obtained from A by differentiating the ith row only.

4.21 Consider the matrix D defined in (2.76), associated with the complex matrix $A = A_1 + iA_2$. Deduce, using the result (b) of Exercise 2.14, that $D^{-1} \sim A^{-1}$.

4.22 Consider the LU decomposition of Section 3.3, as described in (3.32) and (3.33). Prove that $|A_2| = a_{11}\alpha_1$, $|A_3| = a_{11}\alpha_1\alpha_2, \ldots, |A| = a_{11}\alpha_1\alpha_2 \cdots \alpha_{n-1}$. Hence deduce that for L and U to exist and be unique, all the leading principal submatrices A_r of A must be nonsingular.

4.23 Consider the 4×4 tridiagonal matrix A in (3.42), and the product $A = LU$ in (3.43) with L and U bidiagonal. Obtain expressions for L^{-1} and U^{-1}, and hence deduce that in general A^{-1} will have *no* zero elements.

4.24 If $u(t)$ and $v(t)$ are differentiable functions of t and $y(t) = u(t)v(t)$, use the product rule for differentiation to show that

$$[u, \dot{u}, \ddot{u}]A = [y, \dot{y}, \ddot{y}]$$

where A is an upper triangular matrix. Similarly, if $z = u/v$ use the quotient rule for differentiation to show that

$$[u, \dot{u}, \ddot{u}]B = [z, \dot{z}, \ddot{z}]$$

where B is also an upper triangular matrix. Verify that $B = A^{-1}$.

4.25 The expression for a 3×3 determinant in Eq. (4.9) can be obtained as follows:

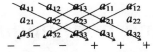

The first two columns of A are repeated on the right of the original array, and products formed along the arrows, with signs attached as shown. Use this method to calculate det$A$ for A in (4.19).

4.26 Write a computer program to evaluate the determinant of square matrices of order five or less using gaussian elimination. Allow for the necessity of row interchanges if there are zero pivots, and for the possibility of a singular matrix. Test your program on Problem 4.16b, and then use it to evaluate

$$\begin{vmatrix} 13 & 39 & 2 & 57 & 28 \\ -4 & -12 & 0 & -19 & -9 \\ 3 & 0 & -9 & 2 & 1 \\ 6 & 17 & 9 & 5 & 7 \\ 19 & 42 & -17 & 107 & 44 \end{vmatrix}$$

5. Rank and non-unique solution of equations

Apart from an informal discussion in Section 3.1 for the case of two equations and unknowns, we have until now assumed that in the equations

$$Ax = b \qquad (5.1)$$

A is square and the solution is unique. However, in many practical applications A may be rectangular – for example, in linear programming problems (Chapter 1) there are usually fewer equations than unknowns. In this chapter we develop the theory necessary to deal with the situation where the solution of (5.1) is not unique, including the case when A is rectangular. There are two problems – first, to determine whether the equations do indeed have a solution (i.e., whether they are *consistent*); and then, if so, to calculate this solution in its most general form. These problems will be solved completely in Section 5.4.

We begin by restating a result from Chapter 4:

5.1 Unique solution

Equations (5.1) have the unique solution $x = A^{-1}b$ if and only if A is a square nonsingular matrix. If $b \equiv 0$ the equations (5.1) are called *homogeneous*, and this unique solution is then $x = A^{-1}0 = 0$, which is called the *trivial* solution. Thus when A is square and $b \equiv 0$, Eqs (5.1) can have a nontrivial solution $x \neq 0$ only if A is singular (i.e., $\det A = 0$), but this solution will not be unique.

Example 5.1 The pair of equations

$$x_1 + x_2 = 0, \qquad x_1 + 2x_2 = 0$$

clearly has only the trivial solution $x_1 = x_2 = 0$. However, if the second equation is replaced by $2x_1 + 2x_2 = 0$ then

$$A = \begin{bmatrix} 1 & 1 \\ 2 & 2 \end{bmatrix} \qquad (5.2)$$

76

is singular, and the equations have the solution $x_1 = t$, $x_2 = -t$, where t is an arbitrary parameter.

Problem 5.1 If x and y are two solutions of (5.1) with A nonsingular, prove directly that $x \equiv y$.

5.2 Definition of rank

A new idea we need is that of the *rank* of an $m \times n$ matrix A. This is denoted by $R(A)$, and is defined to be the order of the largest nonsingular square submatrix which can be formed by selecting rows and columns of A (as we noted in Section 2.4, a submatrix can include A itself).

Example 5.2
(a) The rank of A in Eq. (5.2) is equal to 1, since $\det A = 0$, but $|a_{11}| = 1 \neq 0$.
(b) The rank of

$$A = \begin{bmatrix} 2 & 4 & 8 \\ 1 & 2 & 1 \end{bmatrix} \tag{5.3}$$

is 2, since the submatrix formed by rows 1, 2 and columns 1, 3 is nonsingular, i.e.,

$$\det \begin{bmatrix} 2 & 8 \\ 1 & 1 \end{bmatrix} = 2 - 8 \neq 0.$$

Notice, however, that the submatrix formed by rows 1, 2 and columns 1, 2 is singular.
(c) Consider

$$A = \begin{bmatrix} 1 & 2 & 4 & 1 \\ 2 & 4 & 8 & 2 \\ 3 & 6 & 2 & 0 \end{bmatrix} \tag{5.4}$$

Each of the four 3×3 submatrices obtained by dropping one of the columns of A is singular, since the second row is twice the first row (see Problem 4.8). However, the submatrix formed by rows 1, 3 and columns 1, 3 is nonsingular, i.e.,

$$\det \begin{bmatrix} 1 & 4 \\ 3 & 2 \end{bmatrix} = 2 - 12 \neq 0$$

so $R(A) = 2$.

We note the following important facts about rank:
(a) The only matrices which have zero rank are those having *all* their elements zero.

77

(b) When forming submatrices the rows (and columns) need not be adjacent.
(c) If there is at *least one* $r \times r$ nonsingular submatrix, but all possible larger-order square submatrices are singular, then $R(A)$ is equal to r – it doesn't matter how many such $r \times r$ nonsingular submatrices there are.
(d) If A is $n \times n$ then $R(A) = n$ if and only if A is nonsingular.
(e) If A is $m \times n$ then $R(A)$ cannot exceed the smaller of m and n, i.e., $R(A) \leq \min(m, n)$.

Problem 5.2 For what values of k will the matrix

$$A = \begin{bmatrix} 1 & 2 & 3 & 2 \\ 3 & 6 & 9 & 6 \\ 4 & 8 & 12 & k \end{bmatrix}$$

have (a) $R(A) = 1$, (b) $R(A) = 2$, (c) $R(A) = 3$?

Problem 5.3 Explain why $R(A^T) = R(A)$ for any matrix A.

5.3 Equivalent matrices

The definition of rank may have seemed to the reader somewhat irrelevant to our proclaimed objective of dealing with linear equations. However, we now link the idea of rank with the fundamental tool of gaussian elimination, which we developed in Chapters 3 and 4.

5.3.1 Elementary operations

Let us return to the operations on determinants defined in property PD6, Section 4.1.2:
(a) interchange any two lines of a matrix $[(Li) \leftrightarrow (Lj)]$;
(b) multiply any line by a nonzero scalar k $[(Li) \times k]$;
(c) add an arbitrary multiple of any line to any other line $[(Li) + p(Lj)]$.
These operations applied to rows (or columns) of any matrix A are called *elementary row* (or *column*) *operations*. It is useful to notice how to reverse the effect of an elementary operation: for (a) the two lines are again interchanged; for (b), (Li) is multiplied by $(1/k)$; and the reverse of (c) is $(Li)-p(Lj)$. Thus the inverse of any elementary operation is also an elementary operation of the same type.

The importance of elementary operations for our purposes is the following fact:

Elementary operations do not alter the rank of a matrix.
The justification for this is now given, but the reader who finds the argument difficult to follow can pass on directly to Example 5.3 without loss of continuity. First, recall from Section 4.1.2 that when A is square, property PD6 states that operation (a) alters $\det A$ by a factor -1 and operation (b) alters $\det A$ by a factor k. When A is rectangular it follows that these operations have precisely the same effect on the determinants of *any* square submatrix of A, so singular submatrices remain singular and nonsingular submatrices remain nonsingular (i.e., (a) and (b) do not alter the rank of A). Next, consider elementary operations of type (c) applied to a rectangular matrix A having rank r. By definition all $(r+1) \times (r+1)$ submatrices S_{r+1} of A are singular. After the application of (c), only submatrices containing line i are altered. If S_{r+1} contains both line i *and* line j, then by PD6 its determinant is unaltered, i.e., S_{r+1} remains singular. If S_{r+1} contains only line i, it also remains singular, since on expansion of $\det S_{r+1}$ by the new ith line this determinant is seen to be a sum of minors of order $(r+1)$ of the original matrix; e.g., if $r = 2$, $i = 2$, $j = 4$, then

$$\begin{vmatrix} a_{11} & a_{12} & a_{13} \\ (a_{21}+pa_{41}) & (a_{22}+pa_{42}) & (a_{23}+pa_{43}) \\ a_{31} & a_{32} & a_{33} \end{vmatrix} = \begin{vmatrix} a_{11}a_{12}a_{13} \\ a_{21}a_{22}a_{23} \\ a_{31}a_{32}a_{33} \end{vmatrix} + \begin{vmatrix} a_{11} & a_{12} & a_{13} \\ pa_{41}pa_{42}pa_{43} \\ a_{31} & a_{32} & a_{33} \end{vmatrix}$$

$$= 0 + p \cdot 0$$

since the two determinants on the right-hand side are 3×3 minors of A. Therefore, in all cases S_{r+1} remains singular after application of operation (c), and the same holds for higher-order square submatrices. Thus any operation of type (c) does not increase the rank of A. Furthermore, an operation of type (c) cannot reduce the rank of A; for if it could, then applying the inverse transformation (which is also an operation of type (c)) would bring back the original matrix, and hence increase the rank back to its original value. However, we have just seen that such an increase is impossible. Therefore, operations of type (c) do not change the rank.

Example 5.3 The matrix A in Eq. (5.4) has rank 2. So, therefore, does the matrix obtained from A by applying, for example,

$$(R2) - 2(R1), \ (R3) - 3(R1), \ (R3) \times (-1/10) \tag{5.5}$$

namely,

$$\begin{bmatrix} 1 & 2 & 4 & 1 \\ 0 & 0 & 0 & 0 \\ 0 & 0 & 1 & \dfrac{3}{10} \end{bmatrix} \tag{5.6}$$

To regain (5.4) from (5.6), apply to (5.6):

$$(R3) \times (-10), \ (R3) + 3(R1), \ (R2) + 2(R1) \tag{5.7}$$

The operations (5.7) are the inverses of those in (5.5), but notice that the order in which they are applied is reversed. This has a natural correspondence in real life: for example, to get dressed one puts on a shirt and fastens its buttons; to get undressed the operations are inverted and reversed in order, so first undo the buttons and then take off the shirt! This idea can be used to explain why for matrices the reversal of order occurs in $(AB)^{-1} = B^{-1}A^{-1}$.

Any matrix B obtained by applying elementary transformations to A is said to be *equivalent* to A, since B has the same rank as A; A and B are called *equivalent matrices*. Of course, gaussian elimination involves the application of elementary operations to rows, and therefore does not alter the rank of a matrix. It is thus not surprising that we can now show how to calculate rank using the gaussian procedure, suitably modified.

5.3.2 Calculation of rank

It is clearly not feasible to apply the definition as a practical way of calculating rank except for very simple cases, because a large number of determinants have to be evaluated. Instead, we obtain a simple procedure using elementary operations. The basic idea is to reduce any given matrix to an equivalent matrix whose rank can be determined by inspection.

Consider first the case when A is square. We simply carry out the triangularization using gaussian elimination, described in Section 4.2 for evaluating $\det A$. If all the pivots are nonzero, then $\det A \neq 0$, so $R(A) = n$. If at some stage it is impossible to find a nonzero pivot in a column, then, as we remarked in Section 3.2, the direct gaussian elimination procedure cannot be continued. This means that $\det A = 0$, so $R(A) < n$. The modification required is to interchange the offending *column* with a suitable column to the right. That is, using the notation of (3.17), if at the jth step it is not possible to choose a pivot in column j, i.e., $b_{ij} = 0$, for $i = j, \ j+1, \ldots, n$, then interchange column j with some column k in which there *is* at least one nonzero element to use as a pivot, i.e., $b_{ik} \neq 0$ for some i such that $j \leqslant i \leqslant n$ (see Table 5.1).

A pivot is then selected from this new jth column and the procedure is continued in the usual way. When no further nonzero pivots can be found, $R(A)$ is equal to the total number r of nonzero pivots. This is because the largest nonsingular submatrix of the final reduced triangular form is obtained by taking the first r rows and columns. An example should make this clear.

$$
\begin{array}{cccccc}
b_{11} & \cdots & b_{1j} & \cdot\ \cdot & b_{1k} & \cdot\ \cdot \\
& b_{22} & b_{2j} & \cdot\ \cdot & b_{2k} & \cdot\ \cdot \\
& \cdot & \cdot & \cdot\ \cdot & \cdot & \cdot\ \cdot \\
& \cdot & \cdot & \cdot\ \cdot & \cdot & \cdot\ \cdot \\
& b_{j-1,j-1} & b_{j-1,j} & \cdot\ \cdot & b_{j-1,k} & \cdot\ \cdot \\
0 & 0 & 0 & \cdot\ \cdot & b_{jk} & \\
& 0 & 0 & \cdot\ \cdot & b_{j+1,k} & \\
& \cdot & \cdot & \cdot\ \cdot & \cdot & \\
& \cdot & \cdot & \cdot\ \cdot & \cdot & \\
& 0 & 0 & \cdot\ \cdot & b_{nk} & \\
& & \text{column } j & & \text{column } k &
\end{array}
$$

at least one element is nonzero

Table 5.1 Interchange of columns j and k.

Example 5.4 Suppose that after reducing the first two columns of a 4×4 matrix in the usual way we obtain an equivalent matrix B, which we then transform as follows:

$$
B = \begin{bmatrix} 1 & 2 & 3 & -1 \\ 0 & 2 & 1 & 4 \\ 0 & 0 & 0 & 6 \\ 0 & 0 & 0 & 3 \end{bmatrix} \rightarrow \begin{bmatrix} 1 & 2 & -1 & 3 \\ 0 & 2 & 4 & 1 \\ 0 & 0 & 6 & 0 \\ 0 & 0 & 3 & 0 \end{bmatrix} \quad \text{(C3)} \leftrightarrow \text{(C4)}
$$

$$
\rightarrow \begin{bmatrix} 1 & 2 & -1 & 3 \\ 0 & 2 & 4 & 1 \\ 0 & 0 & 6 & 0 \\ 0 & 0 & 0 & 0 \end{bmatrix} \quad \text{(R4)} - \tfrac{1}{2}\text{(R3)} \qquad (5.8)
$$

No further nonzero pivots can be found in (5.8). Because of the triangular form we see immediately that the largest nonsingular submatrix in (5.8) is formed by the first three rows and columns. Hence $R(A) = 3$, the number of nonzero pivots.

Problem 5.4 Using the triangularization procedure, calculate the rank of

$$
\begin{bmatrix} 1 & 3 & 4 & 6 \\ 2 & 4 & -1 & 2 \\ 3 & 7 & 3 & 8 \\ 5 & 11 & 2 & 10 \end{bmatrix}
$$

The above modified gaussian elimination procedure incorporating column interchanges can be applied in exactly the same way if A is $m \times n$, as is now shown in some further examples.

81

Example 5.5 Using the 3×4 matrix A in Eq. (5.4)

$$A \to \begin{bmatrix} 1 & 2 & 4 & 1 \\ 0 & 0 & 0 & 0 \\ 0 & 0 & -10 & -3 \end{bmatrix} \quad \begin{aligned} &(R2) - 2(R1) \\ &(R3) - 3(R1) \end{aligned}$$

$$\to \begin{bmatrix} 1 & 4 & 2 & 1 \\ 0 & 0 & 0 & 0 \\ 0 & -10 & 0 & -3 \end{bmatrix} \quad (C2) \leftrightarrow (C3)$$

$$\to \begin{bmatrix} 1 & 4 & 2 & 1 \\ 0 & -10 & 0 & -3 \\ 0 & 0 & 0 & 0 \end{bmatrix} \quad (R2) \leftrightarrow (R3) \qquad (5.9)$$

No more nonzero pivots can be found, showing $R(A) = 2$ as before.

Example 5.6 This illustrates the case $m > n$.

$$A = \begin{bmatrix} 1 & 2 & 3 \\ 3 & 6 & 10 \\ 2 & 5 & 7 \\ 1 & 2 & 4 \end{bmatrix} \to \begin{bmatrix} 1 & 2 & 3 \\ 0 & 0 & 1 \\ 0 & 1 & 1 \\ 0 & 0 & 1 \end{bmatrix} \quad \begin{aligned} &(R2) - 3(R1) \\ &(R3) - 2(R1) \\ &(R4) - (R1) \end{aligned}$$

$$\to \begin{bmatrix} 1 & 2 & 3 \\ 0 & 1 & 1 \\ 0 & 0 & 1 \\ 0 & 0 & 1 \end{bmatrix} \quad (R2) \leftrightarrow (R3)$$

$$\to \begin{bmatrix} 1 & 2 & 3 \\ 0 & 1 & 1 \\ 0 & 0 & 1 \\ 0 & 0 & 0 \end{bmatrix} \quad (R4) - (R3) \qquad (5.10)$$

There are three nonzero pivots, so $R(A) = 3$.

Thus, irrespective of whether A is square or rectangular, the nonzero pivots can be put into the $(1,1), (2,2), \ldots, (r,r)$ positions, showing that $R(A) = r$.

Example 5.7

$$A = \begin{bmatrix} 0 & 1 & 3 & 0 & 0 & 6 \\ 0 & 0 & 0 & 0 & 2 & -3 \\ 0 & 0 & 0 & 5 & 0 & 2 \\ 0 & 0 & 0 & 0 & 0 & 1 \end{bmatrix} \to \begin{bmatrix} 1 & 0 & 0 & 6 & 3 & 0 \\ 0 & 2 & 0 & -3 & 0 & 0 \\ 0 & 0 & 5 & 2 & 0 & 0 \\ 0 & 0 & 0 & 1 & 0 & 0 \end{bmatrix}$$

by suitable column interchanges, showing $R(A) = 4$.

Problem 5.5 Determine the rank of the following matrices:

$$(a) \begin{bmatrix} 0 & 1 & 3 & -2 \\ 2 & 2 & -1 & 1 \\ 4 & 5 & 1 & 0 \end{bmatrix} \quad (b) \begin{bmatrix} 3 & 2 & 2 \\ 6 & 4 & 4 \\ -5 & -4 & -6 \\ 10 & 7 & 8 \end{bmatrix}$$

$$(c) \begin{bmatrix} 1 & 0 & -1 & 1 & 0 & 1 \\ 1 & 1 & 3 & 2 & 4 & 3 \\ 2 & 1 & 2 & 3 & 4 & 4 \\ 1 & -2 & -9 & 1 & -8 & -3 \\ 5 & 4 & 11 & 9 & 16 & 13 \end{bmatrix}$$

A useful result which can be conveniently stated here without proof is that if A is $m \times n$ and B is $p \times m$ then

$$R(A) + R(B) - m \leqslant R(BA) \leqslant \min[R(A), R(B)] \tag{5.11}$$

Problem 5.6 Use (5.11) to prove that if $R(A) = m$ then $R(A^T A) = R(A)$ (in fact it can be shown that this result holds whatever the value of $R(A)$).

5.3.3 Normal form

So far as determination of rank is concerned, we have seen that we need to apply only elementary row transformations and column interchanges. It is both interesting and useful to see what further reduction can be obtained by applying additional elementary operations of types (b) and (c) in Section 5.3.1 to the *columns* of the reduced triangular form.

Example 5.8

(a) Consider the matrix in Eq. (5.8). Reduce the rows so as to have zeros to the *right* of the pivots, by applying the following elementary column operations:

$$(C2) - 2(C1), (C3) + (C1), (C4) - 3(C1) \quad \text{(first row)}$$
$$(C3) - 2(C2), (C4) - \tfrac{1}{2}(C2) \quad \text{(second row)}$$

This gives the first matrix below, which can be further reduced as shown

$$\begin{bmatrix} 1 & 0 & 0 & 0 \\ 0 & 2 & 0 & 0 \\ 0 & 0 & 6 & 0 \\ 0 & 0 & 0 & 0 \end{bmatrix} \rightarrow \begin{bmatrix} 1 & 0 & 0 & 0 \\ 0 & 1 & 0 & 0 \\ 0 & 0 & 1 & 0 \\ 0 & 0 & 0 & 0 \end{bmatrix} \quad \begin{matrix} \tfrac{1}{2}(R2) \\ \tfrac{1}{6}(R3) \end{matrix}$$

(b) Similarly, the matrices in (5.9) and (5.10) can be reduced respectively to

$$\begin{bmatrix} 1 & 0 & 0 & 0 \\ 0 & 1 & 0 & 0 \\ 0 & 0 & 0 & 0 \end{bmatrix}, \quad \begin{bmatrix} 1 & 0 & 0 \\ 0 & 1 & 0 \\ 0 & 0 & 1 \\ 0 & 0 & 0 \end{bmatrix}$$

Generally, any $m \times n$ matrix A can be reduced in this fashion by elementary row *and* column operations to an $m \times n$ matrix N having 1's in the $(1, 1), (2, 2), \ldots, (r, r)$ positions and zeros everywhere else, where $r = R(A)$. This special matrix N is called the *normal form* of A, and clearly $R(N) = r = R(A)$.

Problem 5.7 Reduce the three matrices in Problem 5.5 to normal form.

Problem 5.8 Write down the normal forms for matrices having rank 2 and orders 2×2, 2×3, 3×2, 4×3 respectively.

It can be shown that the relationship between A and its normal form can be written

$$A = PNQ \tag{5.12}$$

where P $(m \times m)$ and Q $(n \times n)$ are nonsingular but not unique. We stated earlier that two matrices A and B are equivalent if they have the same dimensions and rank. In this case it is obvious that A and B have the same normal form, so we can write

$$B = P_1 N Q_1 \tag{5.13}$$

with P_1 and Q_1 nonsingular. From (5.13) $N = P_1^{-1} B Q_1^{-1}$ and substituting this into (5.12) gives

$$A = P P_1^{-1} B Q_1^{-1} Q$$
$$= P_2 B Q_2 \tag{5.14}$$

which is an alternative characterization of equivalence of A and B. Notice that P_2 and Q_2 are also nonsingular since each is a product of nonsingular matrices (see Problem 4.18).

The converse result also holds, namely, that if A and B satisfy (5.14) for any nonsingular P_2 and Q_2 then A and B are equivalent. In particular, if T is an arbitrary $m \times m$ nonsingular matrix it follows (by taking $P_2 = T$, $Q_2 = I$) that TA is equivalent to A, i.e., $R(TA) = R(A)$. Similarly, postmultiplication of A by an arbitrary $n \times n$ nonsingular matrix does not alter its rank.

It can also be shown that A and B are equivalent if and only if it is possible to pass from one to the other by a sequence of elementary operations. The proof of this result (and some others in this section, e.g., (5.12)) relies on the introduction of 'elementary matrices' (see Exercises 5.7 and 5.8), but details lie outside the scope of this book. It is interesting to note, however, that the transformation matrices in (5.12) can be easily obtained from the reduction procedure (see Exercise 5.15).

Problem 5.9 For what matrices A are A and A^T equivalent?

Problem 5.10 If A and B are equivalent $n \times n$ matrices, determine whether the following pairs are equivalent: (a) A^T and B^T, (b) A^2 and B^2, (c) AB and BA.
Consider the cases when both A and B are nonsingular, or both are singular.

5.4 Non-unique solution of equations

We have now developed enough theory to tackle the problems stated at the beginning of this chapter.

5.4.1 Homogeneous equations

As pointed out in Section 5.1, the m homogeneous equations

$$Ax = 0 \tag{5.15}$$

in n unknowns $x_1, \ldots, x_n$ have only the trivial solution $x = 0$ if A is square (i.e., $m = n$) and nonsingular. If A is singular or rectangular, express it in normal form as in (5.12). Substituting (5.12) into (5.15) gives $PNQx = 0$, or simply

$$NQx = 0 \tag{5.16}$$

after premultiplying by P^{-1}. Write (5.16) in the form

$$Ny = 0 \tag{5.17}$$

where

$$y = Qx, \quad x = Q^{-1}y \tag{5.18}$$

Recalling that N has r 1's in the $(1,1), \ldots, (r, r)$ positions and zeros elsewhere, where $r = R(A)$, Eqs (5.17) when written out are simply

$$y_1 = 0, \quad y_2 = 0, \quad \ldots, \quad y_r = 0$$

85

MMFE—D

Since the remaining variables $y_{r+1}, \ldots, y_n$ do not appear in the expansion of (5.17) it follows that they can take arbitrary values. Thus the solution x given by (5.18) contains $n - r$ arbitrary parameters.

We have therefore shown that Eqs (5.15) in n unknowns always have a nontrivial solution provided $r = R(A) < n$, and then the most general form of the solution contains $n - r$ arbitrary parameters. If $r = n$, then the only solution is the trivial one, $x = 0$. Notice that if there are fewer equations than unknowns (i.e., $m < n$) then $R(A) \leqslant m < n$, so a nontrivial solution will *always* exist.

The method of determining the solution of (5.15) is to apply the modified gaussian elimination procedure developed in Section 5.3.2 for calculating $R(A)$.

Example 5.9   We find the general solution of three sets of homogeneous equations.

(a) $\quad x_1 + 2x_2 + 4x_3 + \ x_4 = 0$

$\qquad 2x_1 + 4x_2 + 8x_3 + 2x_4 = 0$ $\hfill$ (5.19)

$\qquad 3x_1 + 6x_2 + 2x_3 \qquad = 0$

The matrix A is that given in (5.4), and the reduction was carried out in Example 5.5, producing the final matrix (5.9). It is important to notice that since columns 2 and 3 were interchanged, (5.9) corresponds to the final equations

$$x_1 + \ 4x_3 + 2x_2 + \ x_4 = 0$$
$$- 10x_3 \qquad - 3x_4 = 0 \qquad (5.20)$$

From (5.20) $x_3 = -(3/10)x_4$ and by back substitution

$$x_1 = -2x_2 - 4(-3/10)x_4 - x_4$$
$$= -2x_2 + (1/5)x_4$$

The general solution of Eqs (5.19) is therefore

$$x_1 = -2t_1 + (1/5)t_2, \qquad x_2 = t_1, \qquad x_3 = -(3/10)t_2, \qquad x_4 = t_2 \quad (5.21)$$

where t_1 and t_2 are arbitrary parameters. As expected, since we found $R(A) = 2$ in Example 5.5, there are $4 - 2 = 2$ such parameters.

(b) $\quad x_1 + 2x_2 + \ 3x_3 = 0$

$\qquad 3x_1 + 6x_2 + 10x_3 = 0$

$\qquad 2x_1 + 5x_2 + \ 7x_3 = 0$ $\hfill$ (5.22)

$\qquad x_1 + 2x_2 + \ 4x_3 = 0$

The matrix A was transformed in Example 5.6 into that of Eq. (5.10), showing that $R(A) = 3$. Since $r = n$, Eqs (5.22) have only the trivial solution $x_1 = x_2 = x_3 = 0$.

(c) $\quad x_1 + 2x_2 + 3x_3 - x_4 = 0$

$\qquad x_1 + 4x_2 + 4x_3 + 3x_4 = 0$

$\qquad 2x_1 + 4x_2 + 6x_3 + 4x_4 = 0$ $\hfill$ (5.23)

$\qquad -x_1 - 2x_2 - 3x_3 + 4x_4 = 0$

$$A = \begin{bmatrix} 1 & 2 & 3 & -1 \\ 1 & 4 & 4 & 3 \\ 2 & 4 & 6 & 4 \\ -1 & -2 & -3 & 4 \end{bmatrix} \rightarrow \begin{bmatrix} 1 & 2 & 3 & -1 \\ 0 & 2 & 1 & 4 \\ 0 & 0 & 0 & 6 \\ 0 & 0 & 0 & 3 \end{bmatrix} \quad \begin{array}{l} (R2) - (R1) \\ (R3) - 2(R1) \\ (R4) + (R1) \end{array}$$

$$\rightarrow \begin{array}{cccc} x_1 & x_2 & x_4 & x_3 \\ \begin{bmatrix} 1 & 2 & -1 & 3 \\ 0 & 2 & 4 & 1 \\ 0 & 0 & 6 & 0 \\ 0 & 0 & 3 & 0 \end{bmatrix} \end{array} \quad (C3) \leftrightarrow (C4) \qquad (5.24)$$

It is convenient to record column interchanges by labelling columns with the corresponding variables, as shown in (5.24). The operation $(R4) - \frac{1}{2}(R3)$ on (5.24) then reduces it to the matrix (5.8) in Example 5.4, where it was deduced that $R(A) = 3$. The final equations are

$$x_1 + 2x_2 - x_4 + 3x_3 = 0$$
$$2x_2 + 4x_4 + x_3 = 0$$
$$6x_4 = 0$$

Hence

$$x_4 = 0, \qquad x_2 = -\tfrac{1}{2}x_3, \qquad x_1 = -2(-\tfrac{1}{2}x_3) - 3x_3 = -2x_3$$

The general solution of (5.23) contains $4 - 3 = 1$ arbitrary parameter, and can be written

$$x_1 = -2t_1, \qquad x_2 = -\tfrac{1}{2}t_1, \qquad x_3 = t_1, \qquad x_4 = 0 \qquad (5.25)$$

To summarize, apply gaussian elimination with column interchanges to A until the maximum number $r = R(A)$ of nonzero pivots is obtained in the $(1, 1), \ldots, (r, r)$ positions. If r is equal to n, the number of unknowns, then the only solution is $x = 0$. If $r < n$, the solution is obtained by back substitution using the final reduced form of A, and contains $n - r$ arbitrary parameters. One final point: the actual way in which the parameters are introduced is not unique. For example, in (5.20) we could take $x_4 = -(10/3)x_3$ leading to $x_1 = -2t_1 - (2/3)t_3$, $x_2 = t_1$, $x_3 = t_3$, $x_4 = -(10/3)t_3$ as an alternative *form* of the solution of (5.19); no new values of the x's are obtained, all that has been done is that t_2 in (5.21) has been replaced by $-(10/3)t_3$.

Problem 5.11 Find the general solution of the equations $Ax = 0$ for each of the three matrices A in Problem 5.5.

Problem 5.12 Investigate whether the following equations have a nontrivial solution, and if so determine the solution

(a) $x_1 + 2x_2 - x_3 = 0$
$2x_1 + 4x_2 + 7x_3 = 0$

(b) $x_1 + 2x_2 - 3x_3 = 0$
$2x_1 + 5x_2 + 2x_3 = 0$
$3x_1 - x_2 - 4x_3 = 0$
$7x_1 + 8x_2 - 8x_3 = 0$

5.4.2 Inhomogeneous equations

We now return to the m equations $Ax = b$ in n unknowns, with $b \neq 0$. We shall see that the solution procedure is very similar to the homogeneous case, but first a little care is needed over the question of consistency. To tackle this we again use the normal form (5.12), which when substituted into (5.1) gives $PNQx = b$. Since we defined $y = Qx$ in (5.18), we can write

$$Ny = P^{-1}b \qquad (5.26)$$

Recalling that N has r 1's on the principal diagonal, the set of Eqs (5.26) when written out is

$$y_1 = \beta_1, \quad y_2 = \beta_2, \quad \ldots, \quad y_r = \beta_r, \quad 0 = \beta_{r+1}, \quad \ldots, \quad 0 = \beta_m \quad (5.27)$$

where the β's are the components of the right-hand side vector $P^{-1}b$. Thus for consistency of (5.27), and hence of (5.1), the last $m - r$ components of $P^{-1}b$ must be zero. In practice this means that for consistency, in the final reduced form of the augmented $m \times (n + 1)$ matrix $B = [A, b]$, any row in which the first n elements are all zero must also have the last column element equal to zero. The elimination process is applied exactly as for the homogeneous case, and again since $y_{r+1}, \ldots, y_n$ do not appear in (5.27), we deduce that the general solution contains $n - r$ arbitrary parameters.

Example 5.10 We determine consistency of two sets of equations

(a) $x_1 - 2x_2 + x_3 = -5$
$x_1 + 5x_2 - 7x_3 = 2$
$3x_1 + x_2 - 5x_3 = 1$
$$(5.28)$$

$$B = \begin{bmatrix} 1 & -2 & 1 & \vdots & -5 \\ 1 & 5 & -7 & \vdots & 2 \\ 3 & 1 & -5 & \vdots & 1 \end{bmatrix} \rightarrow \begin{bmatrix} 1 & -2 & 1 & \vdots & -5 \\ 0 & 7 & -8 & \vdots & 7 \\ 0 & 0 & 0 & \vdots & 9 \end{bmatrix} \qquad (5.29)$$

after the usual row operations. The last row of B corresponds to the equation $0x_1 + 0x_2 + 0x_3 = 9$, which cannot be satisfied for any values of the x's, so the Eqs (5.28) are inconsistent.

(b) $2x_1 - 3x_2 + 6x_3 - 5x_4 = 3$

$\qquad x_2 - 4x_3 + x_4 = 1$ $\hfill$ (5.30)

$4x_1 - 5x_2 + 8x_3 - 9x_4 = k$

After the operations (R3) – 2(R1), (R3) – (R2), the augmented matrix becomes

$$\begin{bmatrix} 2 & -3 & 6 & -5 & | & 3 \\ 0 & 1 & -4 & 1 & | & 1 \\ 0 & 0 & 0 & 0 & | & k-7 \end{bmatrix} \qquad (5.31)$$

The last row of (5.31) corresponds to the equation

$$0x_1 + 0x_2 + 0x_3 + 0x_4 = k - 7$$

so Eqs (5.30) are consistent if $k = 7$ and inconsistent if $k \neq 7$.

Problem 5.13 Test for consistency the equations

$$3x_1 - 7x_2 + 14x_3 - 8x_4 = 24$$
$$x_1 - 4x_2 + 3x_3 - x_4 = -2$$
$$x_1 - 3x_2 + 4x_3 - 2x_4 = 4$$
$$2x_1 - 15x_2 - x_3 + 5x_4 = -46$$

Problem 5.14 Find the value of k for which the equations

$$3x_1 + 2x_2 + 4x_3 = 3$$
$$x_1 + x_2 + x_3 = k$$
$$5x_1 + 4x_2 + 6x_3 = 15$$

are consistent.

If the equations are found to be consistent, then the general solution is obtained from the final reduced equations by back substitution, as usual, and contains $n - R(A)$ arbitrary parameters.

Example 5.10 (continued) Consider Eqs (5.30). These were found to be consistent if $k = 7$. In this case the equations corresponding to (5.31) are

$$2x_1 - 3x_2 + 6x_3 - 5x_4 = 3$$
$$x_2 - 4x_3 + x_4 = 1$$

Hence

$$x_2 = 1 + 4x_3 - x_4$$

$$x_1 = \frac{3}{2} + \frac{3}{2}(1 + 4x_3 - x_4) - 3x_3 + \frac{5}{2}x_4$$

$$= 3 + 3x_3 + x_4$$

89

and the general solution can be written

$$x_1 = 3 + 3t_1 + t_2, \qquad x_2 = 1 + 4t_1 - t_2, \qquad x_3 = t_1, \qquad x_4 = t_2 \quad (5.32)$$

where t_1 and t_2 are arbitrary parameters. On inspecting the first four columns of (5.31) we see there are two nonzero pivots, so $R(A) = 2$ and this confirms that the general solution contains $4 - 2 = 2$ arbitrary parameters.

Notice that (5.32) can be written

$$x = [3t_1 + t_2, 4t_1 - t_2, t_1, t_2]^T + [3, 1, 0, 0]^T \qquad (5.33)$$

the sum of a constant vector and one which depends entirely on the parameters. This is true in general: if Eqs (5.1) are consistent then the general solution is

$$x = x_h + x_p$$

where x_h is the general solution of the homogeneous equations in the usual form (5.15) and thus contains no constant terms, and x_p is *any* particular solution of (5.1) and thus is independent of parameters. (The reader who has studied linear differential equations will recognize a parallel with complementary function and particular integral.) For example, in (5.33)

$$x_h = [3t_1 + t_2, 4t_1 - t_2, t_1, t_2]^T$$

is the general solution of the homogeneous equations obtained by replacing the right-hand sides of (5.30) by zeros.

Example 5.11 We find the general solution of two sets of inhomogeneous equations.

(a) $6x_1 + 8x_2 = -1$
   $3x_1 - x_2 = -3$
   $3x_1 + 7x_2 = 1$

After row operations,

$$B = \begin{bmatrix} 6 & 8 & -1 \\ 0 & -5 & -\dfrac{5}{2} \\ 0 & 0 & 0 \end{bmatrix}$$

showing the equations are consistent, and $R(A) = 2 = n$, so the solution is unique. The corresponding equations give $-5x_2 = -5/2$, so $x_2 = \frac{1}{2}$, and $x_1 = (-1 - 8 \cdot \frac{1}{2})/6 = -5/6$. Notice that geometrically, this means that the three straight lines corresponding to the equations intersect at a common point.

(b)
$$x_1 + 2x_2 + 3x_3 - x_4 = -1$$
$$x_1 + 4x_2 + 4x_3 + 3x_4 = 5$$
$$2x_1 + 4x_2 + 6x_3 + 4x_4 = 2$$
$$-x_1 - 2x_2 - 3x_3 + 4x_4 = 3$$

These equations have the same left-hand sides as in (5.23). Applying the same elementary operations as were used in Example 5.9c, but this time on the augmented matrix B, gives (see (5.24))

$$
\begin{array}{cccc}
x_1 & x_2 & x_4 & x_3
\end{array}
$$

$$
\left[\begin{array}{cccc|c}
1 & 2 & -1 & 3 & -1 \\
0 & 2 & 4 & 1 & 6 \\
0 & 0 & 6 & 0 & 4 \\
0 & 0 & 3 & 0 & 2
\end{array}\right]
\rightarrow
\left[\begin{array}{cccc|c}
1 & 2 & -1 & 3 & -1 \\
0 & 2 & 4 & 1 & 6 \\
0 & 0 & 6 & 0 & 4 \\
0 & 0 & 0 & 0 & 0
\end{array}\right]
\quad (R4) - \tfrac{1}{2}(R3)
$$

The equations are consistent, and $r = 3$, $n = 4$. By back substitution

$$x_4 = \frac{2}{3}, \quad x_2 = \frac{1}{2}(6 - 4x_4 - x_3) = \frac{5}{3} - \frac{1}{2}x_3,$$

$$x_1 = -1 - 2x_2 + x_4 - 3x_3 = -\frac{11}{3} - 2x_3$$

and the general solution is

$$x_1 = -\frac{11}{3} - 2t_1, \qquad x_2 = \frac{5}{3} - \frac{1}{2}t_1, \qquad x_3 = t_1, \qquad x_4 = \frac{2}{3}$$

Notice that $x_h = [-2t_1, -\frac{1}{2}t_1, t_1, 0]^T$, which agrees with (5.25), and $x_p = [-\frac{11}{3}, \frac{5}{3}, 0, \frac{2}{3}]^T$.

Problem 5.15 Find the general solution of the equations in Problem 5.13.

Problem 5.16 Find the general solution of the equations in Problem 5.14 when they are consistent.

5.4.3 Consistency theorem

We have seen that for consistency of Eq. (5.1), if any row of the reduced B matrix has all its first n elements equal to zero then the last element in this row must also be zero. This implies that the number of nonzero pivots is the same for B as for A, so $R(B) = R(A)$.

Example 5.12 Consider again the inconsistent equations (5.28). From the reduced form of A in (5.29) (the first three columns) it is clear that

91

$R(A) = 2$, but for B the operation $(C3) \leftrightarrow (C4)$ produces an extra non-zero pivot, showing $R(B) = 3 \neq R(A)$.

The following general theorem (Barnett and Cronin (1975), Section 8.8) can be proved:

For the m equations in n unknowns, $Ax = b$ with augmented matrix $B = [A, b]$, then $R(B) \geq R(A)$ and the equations possess:
(a) a unique solution if and only if $R(A) = R(B) = n$;
(b) an infinite number of solutions if and only if $R(A) = R(B) < n$;
(c) no solution if and only if $R(A) < R(B)$.

This formal result is useful for theoretical purposes, but actual determination of consistency and solutions is carried out as described in Sections 5.4.1 and 5.4.2, using elementary operations.

Problem 5.17 For the equations

$$kx_1 + x_2 + x_3 = 5$$
$$3x_1 + 2x_2 + kx_3 = 18 - 5k$$
$$x_2 + 2x_3 = 2$$

find the value of k: (a) for which there is no solution, (b) for which there are an infinite number of solutions.

Some readers may find a geometrical interpretation of the theorem helpful. The case $n = 2$ was discussed in Examples 3.1 and 3.2 in Section 3.1. When $n = 3$, an equation $a_{11}x_1 + a_{12}x_2 + a_{13}x_3 = b_1$ describes a plane in three dimensions. Thus for three equations, representing three planes, case (a) of the theorem corresponds to the planes intersecting in a single point; case (b) when $R(A) = 2$ corresponds to the planes intersecting in a straight line (like the pages of an opened book); and case (c) corresponds to the planes having no point in common. An illustration of this last case is provided by the three planes in (5.28).

5.5 Method of least squares

It can happen in a number of situations that an inconsistent system of Eqs (5.1) is obtained, but that nevertheless a 'solution' is required which is the 'best possible' in some sense. An important such case is now described.

A common procedure in scientific or other experiments is to measure pairs of values of variables which are associated in some way, and then try to fit some curve to these n points having cartesian coordinates (α_i, β_i), $i = 1, 2, \ldots, n$. A convenient curve is the polynomial expression

$$y = a_0 + a_1 x + a_2 x^2 + \cdots + a_{m-1} x^{m-1} \qquad (5.34)$$

If $m = n$ and the α's are all different, then there is a *unique* curve (5.34) which passes through all the points simultaneously (see Exercise 4.17). Otherwise, if $m < n$ then in general not all the given points can lie on the curve.

The simplest, but none the less very important case is to take $m = 2$, so the curve is simply the straight line

$$y = a_0 + a_1 x \qquad (5.35)$$

and it is required to fit a 'best' line to a set of points as indicated in Fig. 5.1. If it is assumed that the α_i are all different, and are known exactly, then the error e_i at each point is the difference between the value of y given by the line and the measured value β_i, i.e.

$$e_i = a_0 + a_1 \alpha_i - \beta_i, \qquad i = 1, 2, \ldots, n \qquad (5.36)$$

We wish to choose a_0 and a_1 in (5.35) so that the total of these errors is as small as possible. Since the errors e_i may be positive or negative and we are interested in the *net* total, we must consider $|e_i|$, or more conveniently, e_i^2. The method of least squares is to choose a_0 and a_1 so that the sum of the squares of the errors

$$S = \sum_{i=1}^{n} (a_0 + a_1 \alpha_i - \beta_i)^2 \qquad (5.37)$$

is minimized.

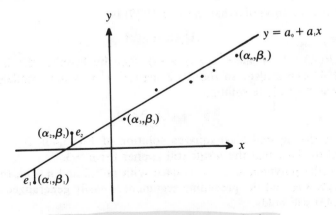

Fig. 5.1 **Best straight line through a set of points.**

93

Notice that if all the points in Fig. 5.1 were to lie on the line we should require

$$a_0 + a_1\alpha_i = \beta_i, \qquad i = 1, 2, \ldots, n \tag{5.38}$$

which represents n equations in the two unknowns a_0 and a_1. Hence if $n > 2$, unless by some freak all the experimentally determined points do indeed lie on the line, this means that Eqs (5.38) are inconsistent.

For simplicity, take $n = 3$ so that (5.37) gives

$$S = (a_0 + a_1\alpha_1 - \beta_1)^2 + (a_0 + a_1\alpha_2 - \beta_2)^2 + (a_0 + a_1\alpha_3 - \beta_3)^2$$

For S to be a minimum each of the partial derivatives $\partial S/\partial a_0$ and $\partial S/\partial a_1$ must be zero. Hence

$$\frac{\partial S}{\partial a_0} = 0 = 2(a_0 + a_1\alpha_1 - \beta_1) + 2(a_0 + a_1\alpha_2 - \beta_2) + 2(a_0 + a_1\alpha_3 - \beta_3)$$

$$\frac{\partial S}{\partial a_1} = 0 = 2\alpha_1(a_0 + a_1\alpha_1 - \beta_1) + 2\alpha_2(a_0 + a_1\alpha_2 - \beta_2) + 2\alpha_3(a_0 + a_1\alpha_3 - \beta_3)$$

and simplifying these two equations gives respectively

$$\left.\begin{array}{r} 3a_0 + (\alpha_1 + \alpha_2 + \alpha_3)a_1 = \beta_1 + \beta_2 + \beta_3 \\ (\alpha_1 + \alpha_2 + \alpha_3)a_0 + (\alpha_1^2 + \alpha_2^2 + \alpha_3^2)a_1 = \alpha_1\beta_1 + \alpha_2\beta_2 + \alpha_3\beta_3 \end{array}\right\} \tag{5.39}$$

If Eqs (5.38) are written in the form

$$Aa = \beta \tag{5.40}$$

where, when $n = 3$,

$$A = \begin{bmatrix} 1 & \alpha_1 \\ 1 & \alpha_2 \\ 1 & \alpha_3 \end{bmatrix}, \qquad a = \begin{bmatrix} a_0 \\ a_1 \end{bmatrix}, \qquad \beta = \begin{bmatrix} \beta_1 \\ \beta_2 \\ \beta_3 \end{bmatrix} \tag{5.41}$$

then it is easy to verify that the pair (5.39) is

$$A^T Aa = A^T\beta \tag{5.42}$$

Since $R(A) = 2$ (because $\alpha_1 \neq \alpha_2 \neq \alpha_3$), then by Problem 5.6 it follows that $R(A^T A) = 2$ also, so the 2×2 matrix $A^T A$ is nonsingular. Hence (5.42) has the unique solution

$$a = (A^T A)^{-1} A^T\beta \tag{5.43}$$

which is the desired least-squares solution of Eq (5.40). It is straightforward to show that the result still applies for $n > 3$.

When the polynomial (5.34) is used with $m > 2$ and $n > m$, then A in (5.40) is $n \times m$ and the preceding argument is easily generalized to show that (5.43) still holds.

Example 5.13 We determine the least-squares quadratic curve for the points having cartesian coordinates $(-1, 3)$, $(0, 0)$, $(1, 2)$, $(2, 5)$.

Here $m = 3$, $n = 4$ and from (5.34) the curve is

$$y = a_0 + a_1 x + a_2 x^2$$

Equations (5.40) are

$$a_0 + a_1 \alpha_i + a_2 \alpha_i^2 = \beta_i, \qquad i = 1, 2, 3, 4$$

which gives

$$A = \begin{bmatrix} 1 & -1 & 1 \\ 1 & 0 & 0 \\ 1 & 1 & 1 \\ 1 & 2 & 4 \end{bmatrix}, \qquad a = \begin{bmatrix} a_0 \\ a_1 \\ a_2 \end{bmatrix}, \qquad \beta = \begin{bmatrix} 3 \\ 0 \\ 2 \\ 5 \end{bmatrix}$$

A simple calculation gives $A^T A$ and $A^T \beta$, and Eqs (5.42) can then be solved by gaussian elimination. The solution is $a = \frac{1}{10}[6, -7, 15]^T$, so the required quadratic is $y = (6 - 7x + 15x^2)/10$. This is shown in Fig. 5.2.

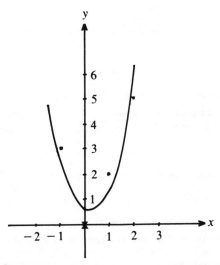

Fig. 5.2 Best quadratic curve for Example 5.13.

Problem 5.18 Use Eq. (5.42) to obtain the best straight line for the points $(1, 2)$, $(2, 4)$, $(3, 5)$, $(4, 7)$. Draw a graph of your result.

Problem 5.19 Verify that when A in (5.40) is square and nonsingular, then (5.43) reduces to the unique solution of the equations (5.40).

A⁻¹ exists iff sq: non (det ≠ 0)

95

(AᵀA)ᵀAa = (Aᵀ)⁻¹ Aᵀ β

I A₂ = I B

Aa = B

5.6 Use of Kronecker product

We now consider a linear equation involving matrices in the form

$$AX = C \tag{5.44}$$

where A is $n \times n$, C is $n \times m$, and the unknown matrix X to be determined is $n \times m$. Of course, if A is nonsingular, we can write the unique solution of (5.44) as $X = A^{-1}C$. However, it is useful to convert (5.44) into the usual matrix–vector form (5.1).

For example, let $n = m = 2$ so (5.44) becomes

$$\begin{bmatrix} a_1 & a_2 \\ a_3 & a_4 \end{bmatrix} \begin{bmatrix} x_1 & x_2 \\ x_3 & x_4 \end{bmatrix} = \begin{bmatrix} c_1 & c_2 \\ c_3 & c_4 \end{bmatrix} \tag{5.45}$$

Expansion and rearrangement of (5.45) easily show that it is equivalent to

$$\begin{bmatrix} a_1 & 0 & a_2 & 0 \\ 0 & a_1 & 0 & a_2 \\ a_3 & 0 & a_4 & 0 \\ 0 & a_3 & 0 & a_4 \end{bmatrix} \begin{bmatrix} x_1 \\ x_2 \\ x_3 \\ x_4 \end{bmatrix} = \begin{bmatrix} c_1 \\ c_2 \\ c_3 \\ c_4 \end{bmatrix} \tag{5.46}$$

and we recognize the 4×4 matrix in (5.46) as $A \otimes I_2$, so (5.46) can be written

$$(A \otimes I_2)x = c$$

where x is the column vector in (5.46) obtained by stacking the elements of the rows of X taken in sequence, and c is formed in the same way from C. It is easy to verify that in general (5.44) can be written

$$(A \otimes I_m)x = c \tag{5.47}$$

where if $X = [x_{ij}]$ then x is the column mn-vector

$$x = [x_{11}, x_{12}, \ldots, x_{1m}, x_{21}, \ldots, x_{2m}, \ldots, x_{nm}]^{\mathrm{T}} \tag{5.48}$$

formed from the rows of X, and similarly for c.

Similarly, if B is an $m \times m$ matrix then it is straightforward to verify that the equation

$$XB = C \tag{5.49}$$

can be written as

$$(I_n \otimes B^{\mathrm{T}})x = c \tag{5.50}$$

where x and c are as in (5.47). What is of most interest is to combine the two expressions (5.47) and (5.50). On doing this it follows that the matrix equation

$$AX + XB = C \tag{5.51}$$

96

(to be solved for X) is equivalent to

$$(A \otimes I_m + I_n \otimes B^T)x = c \tag{5.52}$$

Thus (5.51) has been transformed into (5.52), which represents mn equations in the usual form (5.1). For example, it now follows by the consistency theorem of Section 5.4.3 that the solution X of (5.51) is unique if and only if the $mn \times mn$ matrix

$$D = A \otimes I_m + I_n \otimes B^T \tag{5.53}$$

is nonsingular, in which case the elements of X are given by $x = D^{-1}c$.

A simple criterion for determining nonsingularity of D will be developed in Section 6.3.5. Equation (5.51) arises in a number of applications, and the Kronecker product has again demonstrated its usefulness by converting the problem into a standard set of linear equations.

Problem 5.20 Use (5.52) to solve (5.51) when

$$A = \begin{bmatrix} 1 & 2 \\ 0 & 3 \end{bmatrix}, \qquad B = \begin{bmatrix} 2 & 1 \\ 1 & 4 \end{bmatrix}, \qquad C = I_2$$

Problem 5.21 Use the result in (2.75), Exercise 2 12, to prove that for D in (5.53)

$$D^2 = A^2 \otimes I_m + 2A \otimes B^T + I_n \otimes (B^T)^2 \tag{5.54}$$

5.7 Linear dependence of vectors

Example 5.14 Consider the following three row vectors

$$u = [1, 4], \qquad v = [2, -1], \qquad w = [4, 7] \tag{5.55}$$

It is obvious that $2u + v - w = 0$, and the vectors u, v, w are called *linearly dependent*. Alternatively,

$$w = 2u + v \tag{5.56}$$

and in (5.56) we have written w as a *linear combination* of u and v, i.e.

$$w = x_1 u + x_2 v \tag{5.57}$$

where x_1 and x_2 are scalars. Writing out (5.57) in full:

$$[4, 3] = [x_1, 4x_1] + [2x_2, -x_2]$$
$$= [x_1 + 2x_2, 4x_1 - x_2]$$

So, on equating components

$$x_1 + 2x_2 = 4, \qquad 4x_1 - x_2 = 3$$

or

$$\begin{bmatrix} 1 & 2 \\ 4 & -1 \end{bmatrix} \begin{bmatrix} x_1 \\ x_2 \end{bmatrix} = \begin{bmatrix} 4 \\ 3 \end{bmatrix} \tag{5.58}$$

Thus expressing w as in (5.57) is equivalent to solving a pair of linear equations in the usual form. Notice that the columns of the matrix in (5.58) are u^T and v^T.

These ideas can be generalized: if we have n row m-vectors

$$v_i = [a_{1i}, a_{2i}, \ldots, a_{mi}], \qquad i = 1, 2, \ldots, n \tag{5.59}$$

then expressing $b = [b_1, b_2, \ldots, b_m]$ as a linear combination of the v's, i.e.,

$$x_1 v_1 + x_2 v_2 + \cdots + x_n v_n = b \tag{5.60}$$

is equivalent to the m linear equations in n unknowns

$$Ax = b^T \tag{5.61}$$

where $x = [x_1, \ldots, x_n]^T$ and the columns of the $m \times n$ matrix A are $v_1^T, v_2^T, \ldots, v_n^T$.

The vectors $v_1, v_2, \ldots, v_n$ are *linearly dependent* if it is possible to find a linear combination of them which gives the zero vector, i.e., $b \equiv 0$ in (5.60), so

$$x_1 v_1 + x_2 v_2 + \cdots + x_n v_n = 0 \tag{5.62}$$

with not all the x's zero. If the only solution to (5.62) is the trivial one $x_1 = x_2 = \cdots = x_n = 0$, then the vectors are *linearly independent*. Equations (5.62) are equivalent to the set (5.61) with $b \equiv 0$, which is the set of homogeneous equations studied in Section 5.4.1. Thus the problem of determining linear dependence or independence is equivalent to determining whether or not the homogeneous equations $Ax = 0$ have a nontrivial solution, for which the condition is $R(A) < n$. In particular, if $m < n$ then $R(A) \leq m < n$, so the vectors are *always* linearly dependent.

Example 5.15
(a) $v_1 = [1, 2, 3], \qquad v_2 = [2, 4, 6], \qquad v_3 = [4, 8, 2], \qquad v_4 = [1, 2, 0]$ \qquad (5.63)

Here $m = 3$, $n = 4$ and it is easily seen that (5.62) produces the set of Eqs (5.19) which in matrix form are

$$\begin{bmatrix} 1 & 2 & 4 & 1 \\ 2 & 4 & 8 & 2 \\ 3 & 6 & 2 & 0 \end{bmatrix} \begin{bmatrix} x_1 \\ x_2 \\ x_3 \\ x_4 \end{bmatrix} = 0 \tag{5.64}$$

$$v_1^T \ v_2^T \ v_3^T \ v_4^T$$

In Example 5.9a we found that these equations do have a nontrivial solution, so the vectors in (5.63) are linearly dependent. For example, setting $t_1 = 1$, $t_2 = 0$ in (5.21) gives $x_1 = -2$, $x_2 = 1$, $x_3 = 0$, $x_4 = 0$, so

$$-2v_1 + 1v_2 + 0v_3 + 0v_4 = 0$$

Note that some of the x's in (5.62) can be zero, provided at least one x_i is nonzero.

(b) If we consider only v_1, v_2, and v_3 in (5.63) then (5.62) gives

$$\begin{bmatrix} 1 & 2 & 4 \\ 2 & 4 & 8 \\ 3 & 6 & 2 \end{bmatrix}\begin{bmatrix} x_1 \\ x_2 \\ x_3 \end{bmatrix} = 0 \tag{5.65}$$

It is easily seen that the matrix in (5.65) has rank 2, so a nontrivial solution exists, implying v_1, v_2 and v_3 are linearly dependent.

(c) Finally, consider only v_2 and v_3 in (5.63). Equations (5.62) are now

$$\begin{bmatrix} 2 & 4 \\ 4 & 8 \\ 6 & 2 \end{bmatrix}\begin{bmatrix} x_1 \\ x_2 \end{bmatrix} = 0 \tag{5.66}$$

and since the matrix in (5.66) has rank 2 the only solution of these is $x_1 = x_2 = 0$, showing v_2 and v_3 are linearly independent.

As the preceding example suggests, there is a direct relationship between linear dependence and rank. It can be shown that the rank of *any* matrix A is equal to the maximum number of its columns (regarded as vectors) which are linearly independent. For example, the matrix A in (5.64) has rank 2 (see Example 5.2c) and we have seen in Example 5.15 that at most two of its columns are linearly independent. Since $R(A) = R(A^T)$, and the columns of A^T are the rows of A, it follows that $R(A)$ is also equal to the maximum number of its rows which are linearly independent.

Problem 5.22 For each of the first two matrices in Problem 5.5 find linear combinations of the columns which give zero (use Problem 5.11). Do the same for the rows.

The vectors $v_1, v_2, \ldots, v_n$ in (5.59) are called a *basis* for the set of n-dimensional row vectors if $m = n$ and if they are linearly independent. In this case A is square and has rank n, i.e., A is nonsingular, so the solution of (5.61) is unique. That is, *any* row n-vector b can be expressed as a unique linear combination of the basis vectors. Again, identical remarks apply to a set of n column n-vectors.

Example 5.16 The simplest basis consists of the rows (or columns) of I_n. For example, using again the notation introduced in Problem 2.19, if $n = 3$ then

$$e_1 = [1, 0, 0], \qquad e_2 = [0, 1, 0], \qquad e_3 = [0, 0, 1]$$

form a basis, and if $b = [b_1, b_2, b_3]$ then obviously

$$b = b_1 e_1 + b_2 e_2 + b_3 e_3$$

However, *any* linearly independent set of three vectors will do, e.g.,

$$f_1 = [1, 1, 0], \qquad f_2 = [0, 1, 1], \qquad f_3 = [1, 1, 1]$$

(simply verify that the 3×3 matrix having f_1^T, f_2^T, f_3^T as its columns is nonsingular) and then

$$b = (b_2 - b_3)f_1 + (b_2 - b_1)f_2 + (b_1 - b_2 + b_3)f_3$$

In general, the coefficients in the linear combination of the basis vectors are found by solving (5.61) in the usual way. This idea of interpreting the solution of a set of linear equations in terms of linear relationships between row or column vectors is a very important one for the development of the theory of 'vector spaces', but this lies outside the scope of this book.

Problem 5.23 If $n = 3$ verify that $g_1 = e_1 + e_2 + e_3$, $g_2 = e_1 + 2e_3$, $g_3 = 2e_2 + e_3$ form a basis, and express $b = [b_1, b_2, b_3]$ in terms of this basis.

Exercises

5.1 The idea of controllability, introduced in Example 4.10, can be extended to the case where there are m control variables, so that the set of differential equations (4.36) is replaced by

$$\frac{dx}{dt} = Ax + Bu$$

where B is an $n \times m$ matrix and u is an m-vector. It can be shown that the controllability matrix (4.37) is replaced by the $n \times nm$ matrix

$$\mathscr{C} = [B, AB, A^2 B, \ldots, A^{n-1} B] \tag{5.67}$$

and in this case the condition for controllability is that $R(\mathscr{C}) = n$.
If

$$A = \begin{bmatrix} 1 & 1 & 0 \\ 0 & 1 & 0 \\ 0 & 0 & 1 \end{bmatrix}, \qquad B = \begin{bmatrix} 0 & 1 \\ 1 & 0 \\ 1 & 0 \end{bmatrix}$$

determine whether in this case the system is controllable.

5.2 Suppose that a certain product is stored in two warehouses, in which there are 17 and 11 units of the product available respectively. It is required to deliver the product to three supermarkets, where the requirements are 9,

12, and 7 units respectively. Let x_{ij} be the number of units transported from warehouse number i to supermarket number j ($i = 1, 2; j = 1, 2, 3$). Thus, for example, considering the first supermarket, we must have $x_{11} + x_{21} = 9$, and so on. Write down the remaining four equations which must be satisfied. Show that the total set of equations is consistent and find the general solution.

This is a very simple example of a *transportation problem*, an important special type of LP problem. In practice, if c_{ij} is the unit transport cost from warehouse i to supermarket j, the aim is to find the transport schedule which minimizes the total cost $\Sigma_i \Sigma_j c_{ij} x_{ij}$.

5.3 Prove that for the equations

$$x_1 + 2x_2 + 3x_3 - 3x_4 = k_1$$
$$2x_1 - 5x_2 - 3x_3 + 12x_4 = k_2$$
$$7x_1 + x_2 + 8x_3 + 5x_4 = k_3$$

to be consistent then $37k_1 + 13k_2 - 9k_3 = 0$. Find the general solution when $k_1 = 2$, $k_2 = 4$.

5.4 Determine the values of k such that the equations

$$x_1 + 2x_2 = x_3$$
$$2x_1 + (3 + k)x_2 = 3x_3$$
$$(k - 1)x_1 + 4x_2 = 3x_3$$

have a nontrivial solution. Find the general solution in these cases.

5.5 Find the values of λ for which the equations

$$2x_1 + (4 - \lambda)x_2 + 7 = 0$$
$$(2 - \lambda)x_1 + 2x_2 + 3 = 0$$
$$2x_1 + 5x_2 + 6 - \lambda = 0$$

are consistent. Find the general solution in these cases.

5.6 Find the rank of the matrix

$$A = \begin{bmatrix} 0 & -2 & 4 & 3 \\ 1 & 1 & 1 & 1 \\ 3 & 5 & -1 & a \\ 2 & 1 & 4 & b \end{bmatrix}$$

for all possible values of the parameters a and b. Determine also in each case the general solution of the corresponding equations $Ax = 0$.

5.7 An *elementary matrix* is a matrix obtained from I_n by applying a single elementary operation to it. For example, $(R1) - 2(R2)$ applied to I_3 gives the elementary matrix

$$E = \begin{bmatrix} 1 & -2 & 0 \\ 0 & 1 & 0 \\ 0 & 0 & 1 \end{bmatrix} \tag{5.68}$$

For simplicity consider only elementary row operations. Show that if E is obtained from I_n by any elementary row operation, then forming the product EA produces the matrix A with the same operation applied to its rows (for example, with E in (5.68) then EA is equal to the matrix obtained by applying (R1) − 2(R2) to A).

5.8 The matrix E in (5.68) corresponds to (R1) − 2(R2) applied to I_3; the inverse of this operation is (R1) + 2(R2), which corresponds to the elementary matrix

$$F = \begin{bmatrix} 1 & 2 & 0 \\ 0 & 1 & 0 \\ 0 & 0 & 1 \end{bmatrix}$$

Notice that $F = E^{-1}$; why?

By considering the inverse of each elementary operation (listed in Section 5.3.1), show that in general the inverse of any elementary matrix is itself an elementary matrix.

5.9 If

$$A = \begin{matrix} & m & n & \\ \begin{bmatrix} A_1 & 0 \\ 0 & A_2 \end{bmatrix} & \begin{matrix} p \\ q \end{matrix} \end{matrix}$$

and $R(A_1) = r_1$, $R(A_2) = r_2$, prove that $R(A) = r_1 + r_2$.

5.10 If A is $m \times n$ and B is $n \times m$ with $m < n$, use Eq. (5.11) to prove that BA is singular.

5.11 By using the method which was applied to (5.44), verify that the equation

$$PXQ = R$$

where P is $n \times n$, Q is $m \times m$, X is $n \times m$, and R is $n \times m$, can be rearranged in the form

$$(P \otimes Q^{\mathsf{T}})x = r \qquad (5.69)$$

where x is defined in (5.48) and r is formed in the same way from R.

5.12 Consider the equation

$$A^2X + 2AXB + XB^2 = C \qquad (5.70)$$

where the matrices have the same dimensions as in (5.51). Using (5.47), (5.50), (5.54), and (5.69), show that (5.70) can be written in the form $D^2x = c$, where D is defined in (5.53) and x, c are as in (5.47).

This shows that the solution of (5.70) can be obtained from that of (5.51).

5.13 It can be shown that if A and B are two matrices having the same dimensions then

$$R(A + B) \leqslant R(A) + R(B)$$

Use this result, together with Eq. (5.11), to prove that if X is an idempotent matrix (defined in Exercise 4.14) then $R(X) + R(I_n - X) = n$.

5.14 If X is an idempotent matrix as defined in Exercise 4.14, use Eq. (5.12) and the result (b) of Exercise 2.7 to prove that $R(X) = \text{tr}(X)$.

5.15 Write the relationship (5.12) between an $m \times n$ matrix A and its normal form N as $SAT = N$, where $S = P^{-1}$, $T = Q^{-1}$. Then S and T can be obtained by recording the elementary transformations used to transform A

into N via the array (not a matrix!):

$$\begin{array}{c|c} I_n & T \\ \hline A & I_m \end{array} \rightarrow \begin{array}{c|c} T & \\ \hline N & S \end{array}$$

For example, if

$$A = \begin{bmatrix} 1 & 2 & 3 \\ 2 & 4 & 6 \end{bmatrix}$$

then reduction of A proceeds as follows:

$$\begin{array}{ccc|cc} 1 & 0 & 0 & & \\ 0 & 1 & 0 & & \\ 0 & 0 & 1 & & \\ \hline 1 & 2 & 3 & 1 & 0 \\ 2 & 4 & 6 & 0 & 1 \end{array} \quad \xrightarrow[\text{(R2)}-2\text{(R1)}]{} \quad \begin{array}{ccc|cc} 1 & 0 & 0 & & \\ 0 & 1 & 0 & & \\ 0 & 0 & 1 & & \\ \hline 1 & 2 & 3 & 1 & 0 \\ 0 & 0 & 0 & -2 & 1 \end{array}$$

$$\xrightarrow[\substack{\text{(C2)}-2\text{(C1)} \\ \text{(C3)}-3\text{(C1)}}]{} \quad \begin{array}{ccc|cc} 1 & -2 & -3 & & \\ 0 & 1 & 0 & & \\ 0 & 0 & 1 & & \\ \hline 1 & 0 & 0 & 1 & 0 \\ 0 & 0 & 0 & -2 & 1 \end{array}$$

The row operations on A are also applied to the rows in the lower right block, and the column operations on A are also applied to the upper left block. In this example

$$S = \begin{bmatrix} 1 & 0 \\ -2 & 1 \end{bmatrix}, \quad T = \begin{bmatrix} 1 & -2 & -3 \\ 0 & 1 & 0 \\ 0 & 0 & 1 \end{bmatrix}, \quad N = \begin{bmatrix} 1 & 0 & 0 \\ 0 & 0 & 0 \end{bmatrix}$$

and it is easy to check that $SAT = N$.

Carry out this procedure for the matrix A in Eq. (5.4), and check that your answer is correct.

5.16 For a set of n real row m-vectors $v_1, \ldots, v_n$ it can be shown that they are linearly independent if and only if the $n \times n$ real symmetric *Gram* matrix $G = [g_{ij}]$, with $g_{ij} = v_i^T v_j$, is nonsingular. Use this result to test the vectors in (5.55).

103

6. Eigenvalues and eigenvectors

The underlying problem in much of the theory discussed so far in this book is that of solving sets of linear algebraic equations. To be able to deal with other applications, such as solution of sets of linear differential equations, we need the ideas of this chapter. Indeed, these are essential for virtually all further development of matrices.

6.1 Definitions

The basic problem is to find values of the scalar λ and corresponding column n-vectors x ($\neq 0$) which satisfy the equation

$$Ax = \lambda x \tag{6.1}$$

where A is a given $n \times n$ matrix. Rewriting (6.1) as

$$(\lambda I_n - A)x = 0 \tag{6.2}$$

we see that (6.2) represents a system of homogeneous equations, so, as stated in Section 5.1, for a nontrivial solution x to exist we must have

$$\det(\lambda I_n - A) = 0 \tag{6.3}$$

When the $n \times n$ determinant in (6.3) is expanded, it produces an nth-degree polynomial in λ, called the *characteristic polynomial* $k(\lambda)$ of A, and (6.3) is called the *characteristic equation* of A, i.e.

$$k(\lambda) \equiv \det(\lambda I_n - A) \equiv \lambda^n + k_1 \lambda^{n-1} + k_2 \lambda^{n-2} + \cdots + k_{n-1}\lambda + k_n = 0 \tag{6.4}$$

The n roots $\lambda_1, \lambda_2, \ldots, \lambda_n$ of this equation are called the *eigenvalues* of A (the term *characteristic roots* is also commonly used). When λ in (6.2) is equal to one of the λ_i, then a solution $x \neq 0$ of (6.2) will exist, but as we have seen in Section 5.4.1 this solution will not be unique. Any such vector x satisfying (6.2) when $\lambda = \lambda_i$ is called an *eigenvector* (or *characteristic vector*) corresponding to λ_i. The set $\lambda_1, \lambda_2, \ldots, \lambda_n$ is called the *spectrum* of A. Although the Anglo-German hybrid terms involving 'eigen-' are rather ugly, they are in widespread use and we shall therefore adopt these from now on.

104

Example 6.1 If

$$A = \begin{bmatrix} 1 & 3 \\ 2 & 2 \end{bmatrix} \tag{6.5}$$

then

$$
\begin{aligned}
\det(\lambda I_2 - A) &= \begin{vmatrix} \lambda - 1 & -3 \\ -2 & \lambda - 2 \end{vmatrix} \\
&= (\lambda - 1)(\lambda - 2) - (-3)(-2) \\
&= \lambda^2 - 3\lambda - 4 \\
&= (\lambda + 1)(\lambda - 4)
\end{aligned}
$$

so the eigenvalues of A are $\lambda_1 = -1$, $\lambda_2 = 4$. When $\lambda = -1$ the equations (6.2) are

$$\begin{bmatrix} -2 & -3 \\ -2 & -3 \end{bmatrix} \begin{bmatrix} x_1 \\ x_2 \end{bmatrix} = 0 \tag{6.6}$$

and the solution of (6.6) is easily found to be $x_1 = t_1$, $x_2 = -\frac{2}{3}t_1$, with t_1 an arbitrary parameter. Similarly, when $\lambda = 4$ the solution of (6.2) is $x_1 = t_2$, $x_2 = t_2$. Thus eigenvectors corresponding to the eigenvalues $-1, 4$ respectively have the form

$$u = t_1 \begin{bmatrix} 1 \\ -\frac{2}{3} \end{bmatrix}, \qquad v = t_2 \begin{bmatrix} 1 \\ 1 \end{bmatrix} \tag{6.7}$$

for arbitrary nonzero t_1 and t_2.

As expected (from the theory of homogeneous equations in Chapter 5) the eigenvectors are not unique. However, it is common practice to choose the parameters so that the eigenvectors are *unit vectors*, i.e., they have unit length, where the *length* (or *euclidean norm*) of an n-vector x with real or complex components is defined to be the real non-negative number

$$\|x\| = (|x_1|^2 + |x_2|^2 + \cdots + |x_n|^2)^{1/2} = (x^*x)^{1/2} \tag{6.8}$$

This definition agrees with the usual idea of length of real vectors in two or three dimensions (see (4.35)). When a vector has been converted to a unit vector it is said to be *normalized*. To carry out this normalization, simply divide x by its length $l = \|x\|$, for by the definition (6.8) the length of x/l is

$$
\begin{aligned}
\|x/l\| &= (|x_1|^2/l^2 + |x_2|^2/l^2 + \cdots + |x_n|^2/l^2)^{1/2} \\
&= (|x_1|^2 + |x_2|^2 + \cdots + |x_n|^2)^{1/2}/l \\
&= l/l = 1
\end{aligned}
$$

so x/l is a unit vector.

For example, the lengths of the vectors in (6.7) are respectively

$$l_1 = \left(t_1^2 + \frac{4}{9}t_1^2\right)^{1/2} = \frac{\sqrt{13}}{3}t_1, \qquad l_2 = (t_2^2 + t_2^2)^{1/2} = \sqrt{2}t_2$$

so normalized eigenvectors are u/l_1, v/l_2, i.e.

$$\frac{3}{\sqrt{13}}\begin{bmatrix} 1 \\ -\frac{2}{3} \end{bmatrix}, \qquad \frac{1}{\sqrt{2}}\begin{bmatrix} 1 \\ 1 \end{bmatrix}$$

corresponding to eigenvalues -1 and 4 respectively. Notice that there is an ambiguity over the sign, since $-u/l_1$, $-v/l_2$ are also normalized eigenvectors. In practice, however, this causes no difficulty.

Some further terminology: the vectors x in (6.1) are often called *right* eigenvectors, to distinguish them from *left* eigenvectors y, which are *row* n-vectors satisfying

$$yA = \lambda y \qquad\qquad (6.9)$$

Notice that since (6.9) gives $y(\lambda I_n - A) = 0$, the condition for a nontrivial y to exist is still (6.3), so the values of λ which satisfy (6.9) are exactly the same as those satisfying (6.1). In this book, if no qualification is made then the reader can assume that right eigenvectors are being used. Before studying properties of eigenvalues and eigenvectors, we devote a section to describing a few out of the multitude of applications in which they arise. This will serve to illustrate that (6.1) is definitely not a merely abstract definition.

Problem 6.1 Determine the eigenvalues and associated normalized eigenvectors for the matrix

$$A = \begin{bmatrix} 1 & 3 \\ 3 & 1 \end{bmatrix}$$

Problem 6.2 If $A = \text{diag}[a_{11}, a_{22}, \ldots, a_{nn}]$ show using property PD4 of Section 4.1.2 that $\lambda_i = a_{ii}$ for $i = 1, 2, \ldots, n$. Show also that this result still holds if A is a triangular matrix. Hence give an example of a 3×3 matrix which has all its eigenvalues equal to zero but is not itself a zero matrix.

Problem 6.3 Consider again the 2×2 matrix $A^{(1)}$ in (2.33) associated with the complex number $z_1 = a_1 + ib_1$. Show that the eigenvalues of $A^{(1)}$ are z_1 and $\bar{z}_1$.

Problem 6.4 Prove that for any nonzero scalar c,

$$\|cx\| = |c| \|x\|$$

6.2 Some applications

Example 6.2 Consider a simplified version of the mass–spring system shown in Fig. 1.4, in which it is assumed that the damping forces are negligible, i.e., $d_1 = d_2 = 0$, and that $m_1 = m_2 = 1$. Equations (1.8) and (1.9) can then be written as

$$\begin{bmatrix} \ddot{x}_1 \\ \ddot{x}_2 \end{bmatrix} = -\begin{bmatrix} k_1 & -k_1 \\ -k_1 & (k_1 + k_2) \end{bmatrix}\begin{bmatrix} x_1 \\ x_2 \end{bmatrix}$$

or in matrix form

$$\ddot{x} = -Ax \tag{6.10}$$

where in (6.10), $x = [x_1, x_2]^T$. Intuitively we feel that the motion has an oscillatory nature, which suggests we try

$$x = y \sin \omega t \tag{6.11}$$

as a solution of (6.10), where y is a constant vector. Then differentiating (6.11) gives $\ddot{x} = -\omega^2 y \sin \omega t$, which on substitution into (6.10) produces

$$-\omega^2 y \sin \omega t = -Ay \sin \omega t \tag{6.12}$$

The parameter ω represents the frequency of oscillations, so we reject $\omega = 0$ in (6.12), which then reduces to

$$(\omega^2 I - A)y = 0$$

which has precisely the form (6.2) with $\lambda = \omega^2$. Thus the solution of the set of second-order linear differential equations (6.10) depends on the eigenvalues and eigenvectors of A, and the eigenvalues determine the frequencies of the oscillations.

Example 6.3 As was done in (1.10), we can convert the second-order equations (6.10) into a first-order set by defining new variables $x_3 = \dot{x}_1$, $x_4 = \dot{x}_2$, to obtain

$$\dot{x} = Bx \tag{6.13}$$

where B is the 4×4 matrix obtained from (1.10) by setting $d_1 = d_2 = 0$, $m_1 = m_2 = 1$, and now $x = [x_1, x_2, x_3, x_4]^T$. An approach to solving (6.13) is suggested by the *scalar* differential equation

$$\dot{z} = bz \tag{6.14}$$

On writing dz/dt for $\dot{z}$ and separating the variables, the reader should confirm that the solution of (6.14) is

$$z(t) = e^{bt}\alpha \tag{6.15}$$

where α is a constant. It therefore seems worthwhile to use the vector

$$x(t) = e^{\lambda t}v \tag{6.16}$$

as a trial solution of (6.14), where v is a constant vector and λ a parameter to be determined. Differentiation of (6.16) gives $\dot{x} = \lambda\, e^{\lambda t}v$, and substituting into (6.13):

$$\lambda\, e^{\lambda t}v = B\, e^{\lambda t}v$$

whence, since $e^{\lambda t} \neq 0$,

$$(\lambda I - B)v = 0 \qquad\qquad (6.17)$$

Again we have obtained an equation in the form (6.2), this time showing that the solution of the set of first-order linear differential equations (6.13) depends on the eigenvalues and eigenvectors of B.

Example 6.4 A stretched string has its ends A and B fixed, and four equal particles m are attached at equal distances d apart, the whole resting on a smooth horizontal table, as shown in Fig. 6.1. The tension T

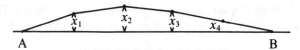

Fig. 6.1   Stretched string for Example 6.4.

in the string is assumed large so that the portions of the string can be assumed straight, and the angles made with AB assumed small. If the displacements of the particles are x_1, x_2, x_3, x_4, it can be shown that the equations of motion are

$$\begin{bmatrix} \ddot{x}_1 \\ \ddot{x}_2 \\ \ddot{x}_3 \\ \ddot{x}_4 \end{bmatrix} = -\frac{T}{md} \begin{bmatrix} 2 & -1 & 0 & 0 \\ -1 & 2 & -1 & 0 \\ 0 & -1 & 2 & -1 \\ 0 & 0 & -1 & 2 \end{bmatrix} \begin{bmatrix} x_1 \\ x_2 \\ x_3 \\ x_4 \end{bmatrix} \qquad (6.18)$$

which can be written in the matrix form (6.10), so the same conclusion about dependence of the solution on eigenvalues and eigenvectors can be drawn. (In passing, notice that the matrix in (6.18) is tridiagonal, as defined in Exercise 3.2.)

The preceding examples illustrate the general fact that the solution of a set of linear differential equations can be expressed in terms of eigenvalues and eigenvectors. Such equations arise in many areas, including mechanical vibrations, electrical networks, and control systems, and will be studied further in Section 6.5 together with linear difference equations, which exhibit similar properties.

Example 6.5 A geometrical interpretation in two and three dimensions can be given. Let P be an arbitrary point in two dimensions having

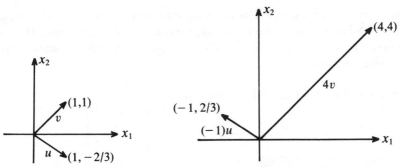

Fig. 6.2a Eigenvectors of A in Eq. (6.5).   Fig. 6.2b Eigenvectors after transformation.

cartesian coordinates x_1 and x_2. If we regard a given 2×2 matrix A as the matrix of a transformation of coordinates (see Example 1.6), then after the transformation has been applied P moves to some point P' having coordinates

$$x' = \begin{bmatrix} x_1' \\ x_2' \end{bmatrix} = A \begin{bmatrix} x_1 \\ x_2 \end{bmatrix} = Ax$$

Thus the condition (6.1) shows that $x' = Ax = \lambda x$. This implies that x' is a scalar multiple of x, which is equivalent to requiring that P' lies on the straight line OP, and its distance from the origin is $OP' = \|x'\| = \|\lambda x\| = |\lambda| \, \|x\| = |\lambda|(OP)$ (see Problem 6.4).

The eigenvectors of A can therefore be thought of as those vectors which are left unaltered in direction after the transformation has been applied, and the eigenvalues measure the ratio of the lengths of these vectors before and after the transformation. For example, the eigenvectors u and v in (6.7) with $t_1 = t_2 = 1$ for the matrix A in (6.5) are shown before and after transformation in Fig. 6.2a and Fig. 6.2b respectively.

Problem 6.5 Deduce that for the transformation in Exercise 1.5a, normalized eigenvectors are unit vectors along the two coordinate axes.

6.3 Properties

6.3.1 The characteristic equation

If A is a real matrix then obviously all the coefficients k_i in the characteristic equation (6.4) will be real, but some of the roots λ_i may be complex. From a well-known result in elementary algebra, any complex roots of (6.4) will occur in conjugate pairs. Eigenvectors corresponding to real eigenvalues will have real elements because the equations (6.2) will then have all coefficients real; however, eigenvectors associated with complex eigenvalues may also have complex elements. If A is a complex matrix then complex eigenvalues will not in general form conjugate pairs.

Problem 6.6 Calculate the eigenvalues of:

$$\text{(a)} \begin{bmatrix} 1 & -10 \\ 3 & -5 \end{bmatrix} \qquad \text{(b)} \begin{bmatrix} 0 & -1 & 1 \\ 2 & 3 & 3 \\ -2 & 1 & 1 \end{bmatrix}$$

Since the roots of (6.4) are $\lambda_1, \ldots, \lambda_n$, we have

$$\det(\lambda I_n - A) \equiv \lambda^n + k_1 \lambda^{n-1} + \cdots + k_n \equiv (\lambda - \lambda_1)(\lambda - \lambda_2) \cdots (\lambda - \lambda_n) \quad (6.19)$$

Setting $\lambda = 0$ in the identity (6.19) gives

$$\det(-A) = k_n = (-\lambda_1)(-\lambda_2) \cdots (-\lambda_n) \quad (6.20)$$

It follows from property PD2 of Section 4.1.2 that $\det(-A) = (-1)^n \det A$, so the expressions in (6.20) give

$$\det A = \lambda_1 \lambda_2 \cdots \lambda_n = (-1)^n k_n \quad (6.21)$$

The result expressed by (6.21) is an interesting one: it shows that the determinant of A is equal to the product of the eigenvalues of A; thus in particular A is singular if and only if it has at least one zero eigenvalue. Next consider the coefficients of λ^{n-1} in (6.19). For simplicity take the case $n = 3$. The right-hand side of (6.19) is then

$$(\lambda - \lambda_1)(\lambda - \lambda_2)(\lambda - \lambda_3) = \lambda^3 - \lambda^2(\lambda_1 + \lambda_2 + \lambda_3) + \cdots$$

and the left-hand side of (6.19) is

$$\det(\lambda I_3 - A) = \begin{vmatrix} \lambda - a_{11} & -a_{12} & -a_{13} \\ -a_{21} & \lambda - a_{22} & -a_{23} \\ -a_{31} & -a_{32} & \lambda - a_{33} \end{vmatrix} \quad (6.22)$$

$$= (\lambda - a_{11})A_{11} + (-a_{12})A_{12} + (-a_{13})A_{13}$$

on expanding by the first row using Eq. (4.23) (the A_{1i} are here the cofactors of the first row of $\lambda I - A$). The only term involving λ^2 in (6.22) is contained in

$$(\lambda - a_{11})A_{11} = (\lambda - a_{11}) \begin{vmatrix} \lambda - a_{22} & -a_{23} \\ -a_{32} & \lambda - a_{33} \end{vmatrix}$$

$$= (\lambda - a_{11})(\lambda^2 - a_{22}\lambda - a_{33}\lambda + \cdots$$

$$= \lambda^3 - (a_{11} + a_{22} + a_{33})\lambda^2 + \cdots$$

Hence when $n = 3$, comparing the coefficients of λ^2 in (6.19) gives

$$-(a_{11} + a_{22} + a_{33}) = k_1 = -(\lambda_1 + \lambda_2 + \lambda_3)$$

showing that the sum of the eigenvalues of A is equal to the sum of the elements on the principal diagonal, i.e., the *trace* of A. A similar argument applies for the general case, giving

$$\lambda_1 + \lambda_2 + \cdots + \lambda_n = \text{tr}(A) = -k_1 \tag{6.23}$$

For example, for the matrix A in (6.5), $\text{tr}(A) = 1 + 2 = 3$ and $\lambda_1 + \lambda_2 = -1 + 4 = 3$.

Expressions involving minors of A can be derived for the other coefficients $k_2, \ldots, k_{n-1}$ in the characteristic equation, but these are more complicated. In fact when n is large it is difficult to calculate the characteristic equation and practical procedures for determining eigenvalues avoid this (see Section 6.6). For 2×2 and 3×3 matrices we can find $\det(\lambda I - A)$ using the cofactor formulae (4.23) or (4.24).

Example 6.6 The characteristic polynomial of the matrix

$$A = \begin{bmatrix} -1 & 0 & 2 \\ 0 & 1 & 2 \\ 2 & 2 & 0 \end{bmatrix} \tag{6.24}$$

is

$$\det(\lambda I_3 - A) = \begin{vmatrix} \lambda+1 & 0 & -2 \\ 0 & \lambda-1 & -2 \\ -2 & -2 & \lambda \end{vmatrix} \tag{6.25}$$
$$= (\lambda+1)[(\lambda-1)\lambda-4] - 2 \cdot 2(\lambda-1)$$
$$= \lambda^3 - 9\lambda = \lambda(\lambda-3)(\lambda+3)$$

(expanding the determinant by the first row). Thus the eigenvalues of A are $\lambda_1 = 0$, $\lambda_2 = 3$, $\lambda_3 = -3$. To obtain an eigenvector corresponding to λ_1, set $\lambda = 0$ in the right-hand side of (6.25) so as to obtain the coefficients in the equations (6.2), which are therefore

$$x_1 \quad\quad -2x_3 = 0$$
$$-x_2 - 2x_3 = 0$$
$$-2x_1 - 2x_2 \quad\quad = 0$$

It is easily confirmed that the general solution of these equations is $x_1 = 2t_1$, $x_2 = -2t_1$, $x_3 = t_1$ for arbitrary nonzero t_1. Thus an eigenvector of A corresponding to λ_1 is $x = t_1[2, -2, 1]^T$. Since by (6.8), $\|x\| = (4t_1^2 + 4t_1^2 + t_1^2)^{1/2} = 3t_1$, a normalized eigenvector corresponding to λ_1 is $x/3t_1 = \frac{1}{3}[2, -2, 1]^T$.

In general, the set of homogeneous equations (6.2) in the n unknowns $x_1, \ldots, x_n$ is solved for each value λ_i of λ by the method of Section

5.4.1. Since by construction $\lambda_i I_n - A$ is singular, these equations will always have a nontrivial (and non-unique) solution. If all the λ_i are all different from each other, then $R(\lambda_i I_n - A) = n - 1$, and each eigenvector will contain just one arbitrary parameter.

Problem 6.7 Determine normalized eigenvectors: (a) for the eigenvalues λ_2 and λ_3 of A in (6.24); (b) for the matrix in Problem 6.6b.

Problem 6.8 For each of the following matrices A expand $\det(\lambda I_3 - A)$ using cofactors, and hence calculate the characteristic equation and eigenvalues:

$$\text{(a)} \begin{bmatrix} -2 & -1 & 0 \\ 1 & 2 & 3 \\ 4 & 5 & 6 \end{bmatrix} \quad \text{(b)} \begin{bmatrix} 1 & 0 & -1 \\ 1 & 2 & 1 \\ 2 & 2 & 3 \end{bmatrix}$$

6.3.2 Hermitian and symmetric matrices

It was remarked in Section 6.3.1 that even real matrices may in general possess some complex eigenvalues (and eigenvectors). However, an important exception to this is provided by hermitian and real symmetric matrices (defined in Section 2.3.2) which always have all their eigenvalues real.

To prove this, let A be an $n \times n$ hermitian matrix, so from the definition (2.56)

$$A^* = A \tag{6.26}$$

and let λ and u be associated eigenvalue and eigenvector of A, so that

$$Au = \lambda u \tag{6.27}$$

At this stage we cannot assume either λ or u is real, so if we take the complex conjugate of both sides of (6.27) and then transpose, we obtain

$$u^* A^* = \bar{\lambda} u^* \tag{6.28}$$

Postmultiplying (6.28) by u and applying (6.26) results in

$$u^* A u = \bar{\lambda} u^* u \tag{6.29}$$

However, premultiplying (6.27) by u^* gives

$$u^* A u = \lambda u^* u$$

and subtracting (6.29) from this last equation produces

$$(\lambda - \bar{\lambda}) u^* u = 0 \tag{6.30}$$

Now u^*u is the scalar product of $u^* = [\bar{u}_1, \ldots, \bar{u}_n]$ and u, which by (2.44) is

$$u^*u = \bar{u}_1 u_1 + \bar{u}_2 u_2 + \cdots + \bar{u}_n u_n$$
$$= |u_1|^2 + |u_2|^2 + \cdots + |u_n|^2 = \|u\|^2 \tag{6.31}$$

and the expression in (6.31) is nonzero since $u \neq 0$. Hence we can conclude from (6.30) that $\bar{\lambda} = \lambda$, showing that λ is real, as required. Since a real symmetric matrix is the special case of a hermitian matrix when all the elements are real, the result also applies in this case. In fact we can also see that when A in (6.27) is real and symmetric, then since all the numbers occurring in (6.27) will be real the eigenvectors will also be real.

Problem 6.9 Calculate the eigenvalues and corresponding normalized eigenvectors for the hermitian matrix

$$A = \begin{bmatrix} 3 & 1+i \\ 1-i & 2 \end{bmatrix}$$

Problem 6.10 Prove that the determinant of any hermitian matrix is a real number.

Problem 6.11 If A is a skew hermitian matrix (defined in (2.57)) show that all its eigenvalues have zero real part.

6.3.3 Matrix polynomials and the Cayley–Hamilton theorem

We now show how to obtain eigenvalues and eigenvectors of matrices derived from A. Let λ_i and u_i be associated eigenvalues and right eigenvectors of an arbitrary $n \times n$ matrix A, so that

$$Au_i = \lambda_i u_i, \qquad i = 1, 2, \ldots, n \tag{6.32}$$

Notice that we are here using u_i to stand for a column n-vector rather than an ith component – it will always be clear from the context when this is meant. Premultiply both sides of (6.32) by A to give

$$A^2 u_i = \lambda_i A u_i = \lambda_i (\lambda_i u_i) = \lambda_i^2 u_i$$

which shows that A^2 has eigenvalues λ_i^2 and eigenvectors u_i, for $i = 1, 2, \ldots, n$. Repeated premultiplication by A similarly shows that A^m has eigenvalues λ_i^m and eigenvectors u_i for any positive integer m, i.e.,

$$A^m u_i = \lambda_i^m u_i \tag{6.33}$$

A slightly different argument shows that A^T has the same eigenvalues as those of A. For we have

$$\det(\lambda I - A^T) = \det(\lambda I - A^T)^T, \quad \text{by property PD1, Section 4.1.2}$$
$$= \det(\lambda I - A), \quad \text{by (2.38)}$$

showing that the characteristic equation of A^T is identical to that of A.

Problem 6.12 Show that for u_i defined in (6.32), u_i^T is a left eigenvector of A^T.

Problem 6.13 For A in (6.32), prove: (a) the eigenvalues of A^* are $\bar{\lambda}_i$; (b) if A is nonsingular, the eigenvalues and eigenvectors of A^{-1} are $1/\lambda_i$ and u_i for $i = 1, 2, \ldots, n$. What are the eigenvalues of $\text{adj}A$?

Problem 6.14 Let A and B be two arbitrary $n \times n$ matrices and let the eigenvalues and eigenvectors of AB be μ_i and v_i, $i = 1, 2, \ldots, n$. Prove that the eigenvalues and eigenvectors of BA are μ_i and Bv_i.

We now generalize the result in (6.33). Let

$$p(\lambda) = \lambda^r + p_1 \lambda^{r-1} + \cdots + p_{r-1} \lambda + p_r \tag{6.34}$$

be an arbitrary polynomial of degree r. Then we define the matrix *polynomial* $p(A)$ in the matrix A to be the $n \times n$ matrix obtained by replacing λ by A, i.e.,

$$p(A) = A^r + p_1 A^{r-1} + \cdots + p_{r-1} A + p_r I_n \tag{6.35}$$

If u_i is defined by (6.32) then

$$p(A)u_i = A^r u_i + p_1 A^{r-1} u_i + \cdots + p_{r-1} A u_i + p_r u_i$$
$$= \lambda_i^r u_i + p_1 \lambda_i^{r-1} u_i + \cdots + p_{r-1} \lambda_i u_i + p_r u_i, \quad \text{by (6.33)}$$
$$= p(\lambda_i) u_i$$

showing that the eigenvalues and eigenvectors of $p(A)$ are $p(\lambda_i)$ and u_i, for $i = 1, 2, \ldots, n$.

Problem 6.15 For A in (6.32), if c is a scalar prove that the eigenvalues of $A + cI_n$ are $\lambda_i + c$, $i = 1, 2, \ldots, n$: (a) by using the preceding result; (b) by using the definition (6.3). Prove also that if $\lambda_i + c \neq 0$ for all i, then the eigenvalues and eigenvectors of $(A + cI)^{-1}$ are $1/(\lambda_i + c)$ and u_i.

Problem 6.16 Prove that $p(A)$ commutes with A for any polynomial $p(\lambda)$ and any square matrix A.

Notice that if λ_i is a root of $p(\lambda)$ then the preceding argument shows that $p(A)u_i \equiv 0$. In particular, a most important result arises when we take $p(\lambda)$ to be the characteristic polynomial of A defined in (6.4), i.e.,

$$k(\lambda) = \det(\lambda I_n - A) = \lambda^n + k_1\lambda^{n-1} + \cdots + k_{n-1}\lambda + k_n \qquad (6.36)$$

We shall prove:

Cayley–Hamilton theorem
Every matrix satisfies its own characteristic equation, i.e.,

$$k(A) \equiv A^n + k_1 A^{n-1} + \cdots + k_{n-1}A + k_n I_n \equiv 0 \qquad (6.37)$$

Example 6.7 The characteristic polynomial of A in Eq. (6.5) was found to be $\lambda^2 - 3\lambda - 4$, and

$$A^2 - 3A - 4I_2 = \begin{bmatrix} 7 & 9 \\ 6 & 10 \end{bmatrix} - 3\begin{bmatrix} 1 & 3 \\ 2 & 2 \end{bmatrix} - 4\begin{bmatrix} 1 & 0 \\ 0 & 1 \end{bmatrix}$$
$$= \begin{bmatrix} 0 & 0 \\ 0 & 0 \end{bmatrix}$$

thereby verifying (6.37) in this case.

It must be stressed that (6.37) is *not* an obvious result: the equation $k(\lambda) = 0$ is satisfied only for the n values $\lambda = \lambda_1, \lambda_2, \ldots, \lambda_n$, whereas (6.37) states that the matrix polynomial $k(A)$ is identically equal to the $n \times n$ zero matrix.

For ease of understanding we give a proof of the Cayley–Hamilton theorem for $n = 3$, but the method generalizes directly for any value of n. First, from (4.46) we have

$$(\lambda I_3 - A)^{-1} = \operatorname{adj}(\lambda I_3 - A)/\det(\lambda I_3 - A)$$
$$= \operatorname{adj}(\lambda I - A)/k(\lambda) \qquad (6.38)$$

where for convenience the suffix on I has been dropped. Recall next that $\operatorname{adj}(\lambda I - A)$ is the transposed matrix of cofactors of $\lambda I - A$, so each of its elements is a 2×2 determinant which will be a polynomial in λ of degree at most 2. For example, from the array in (6.22) we see that the 2,2 element of $\operatorname{adj}(\lambda I - A)$ is

$$\begin{vmatrix} \lambda - a_{11} & -a_{13} \\ -a_{31} & \lambda - a_{33} \end{vmatrix} = (\lambda - a_{11})(\lambda - a_{33}) - a_{13}a_{31}$$

Collecting together powers of λ, we can therefore write

$$\operatorname{adj}(\lambda I - A) = \lambda^2 B_1 + \lambda B_2 + B_3 \qquad (6.39)$$

where B_1, B_2, and B_3 are constant 3×3 matrices. Combining together (6.38) and (6.39) gives

$$I = (\lambda I - A)(\lambda I - A)^{-1}$$
$$= (\lambda I - A)(\lambda^2 B_1 + \lambda B_2 + B_3)/k(\lambda) \tag{6.40}$$

and multiplying both sides of (6.40) by $k(\lambda)$ produces

$$(\lambda^3 + k_1\lambda^2 + k_2\lambda + k_3)I = (\lambda I - A)(\lambda^2 B_1 + \lambda B_2 + B_3)$$
$$= \lambda^3 B_1 + \lambda^2(B_2 - AB_1) + \lambda(B_3 - AB_2) - AB_3 \tag{6.41}$$

Equating coefficients of powers of λ in (6.41):

$$\begin{aligned} \lambda^3: & \quad I \equiv B_1 \\ \lambda^2: & \quad k_1 I \equiv B_2 - AB_1 \\ \lambda: & \quad k_2 I \equiv B_3 - AB_2 \\ \lambda^0: & \quad k_3 I \equiv -AB_3 \end{aligned} \tag{6.42}$$

Finally, premultiply the identities in (6.42) by A^3, A^2, A, and I respectively, and add the resulting expressions to obtain

$$A^3 + k_1 A^2 + k_2 A + k_3 I \equiv A^3 B_1 + (A^2 B_2 - A^3 B_1) + (AB_3 - A^2 B_2) - AB_3$$
$$\equiv 0$$

which is the desired result. The only differences for $n > 3$ are that (6.39) is replaced by $\lambda^{n-1}B_1 + \cdots + B_n$, and there are $n + 1$ identities like (6.42), but otherwise the proof proceeds in the same way.

Problem 6.17 Verify the Cayley–Hamilton theorem for the two matrices in Problem 6.8 (i.e., evaluate A^3 and A^2 and substitute into the characteristic polynomial).

Problem 6.18 Obtain the characteristic equation of the 3×3 matrix A in Eq. (2.25), and hence write down (6.37) in this case. Notice that it was verified in Problem 2.9 that for this matrix A we have $A^2 - 3A + 2I_3 \equiv 0$.

Problem 6.18 illustrates that it is possible for some $n \times n$ matrices A to have $q(A) \equiv 0$, where $q(\lambda)$ is a polynomial having degree *less* than n. Such a matrix is called *derogatory*. The polynomial $m(\lambda)$ of lowest degree such that $m(A) \equiv 0$ is called the *minimum polynomial* of A. When $m(\lambda)$ is the same as the characteristic polynomial $k(\lambda)$, then A is called *nonderogatory*. An important condition which ensures that A is nonderogatory is that all its eigenvalues are *distinct* (i.e., all different from each other). However, not all matrices possessing repeated eigenvalues are derogatory.

Equation (6.37) can be rearranged to give

$$A^n = -k_1 A^{n-1} - k_2 A^{n-2} - \cdots - k_{n-1}A - k_n I \qquad (6.43)$$

which expresses A^n as a linear combination of $A^{n-1}, A^{n-2}, \ldots, A, I$. If (6.43) is premultiplied by A, we get A^{n+1} expressed as a linear combination of $A^n, A^{n-1}, \ldots, A, I$, and hence it follows that A^{n+1} can be expressed as a linear combination of $A^{n-1}, A^{n-2}, \ldots, I$. Similarly, by repeated premultiplication of (6.43) by A it follows that *any* power $A^{n+t}, t = 0, 1, 2, \ldots$, can be expressed as a linear combination of powers of A up to A^{n-1}.

Example 6.8 Consider again the 2×2 matrix A in (6.5). In Example 6.7 we found that

$$A^2 = 3A + 4I \qquad (6.44)$$

Hence

$$A^3 = 3A^2 + 4A = 3(3A + 4I) + 4A = 13A + 12I$$
$$= \begin{bmatrix} 25 & 39 \\ 26 & 38 \end{bmatrix}$$

and similarly

$$A^4 = 13A^2 + 12A = 13(3A + 4I) + 12A = 51A + 52I$$

and so on.

On the other hand, if A is nonsingular, we can premultiply (6.37) by A^{-1} and rearrange to get

$$k_n A^{-1} = -(A^{n-1} + k_1 A^{n-2} + \cdots + k_{n-1}I) \qquad (6.45)$$

Since A is nonsingular, (6.21) shows that $k_n \neq 0$, so we can divide in (6.45) to give

$$A^{-1} = -(A^{n-1} + k_1 A^{n-2} + \cdots + k_{n-1}I)/k_n \qquad (6.46)$$

Example 6.8 (continued) Premultiplying (6.44) by A^{-1} gives

$$A = 3I + 4A^{-1}$$

so that

$$A^{-1} = \frac{1}{4}(A - 3I) = \frac{1}{4}\begin{bmatrix} -2 & 3 \\ 2 & -1 \end{bmatrix}$$

In general, calculation of A^{-1} via (6.46) is not recommended because of the previously mentioned difficulty of calculating the k's.

Problem 6.19 Determine A^4 and A^{-1} for the matrix in Problem 6.8b using the Cayley–Hamilton theorem.

117

MMFE—E

The use of the Cayley–Hamilton theorem to evaluate powers of A can be extended to evaluation of $p(A)$, where $p(\lambda)$ is an arbitrary polynomial (6.34) having degree *greater* than n. To see how this is done, first recall the fact from elementary algebra that if $p(\lambda)$ is divided by $k(\lambda)$ to give some quotient $q(\lambda)$, say, then the remainder $r(\lambda)$ has degree *less* than n, i.e.,

$$p(\lambda) \equiv q(\lambda)k(\lambda) + r(\lambda) \tag{6.47}$$

Now (6.47) is an identity, not an equation, so it still holds if we replace λ by A, which gives

$$p(A) \equiv q(A)k(A) + r(A) \tag{6.48}$$

However, by the Cayley–Hamilton theorem $k(A) \equiv 0$, so (6.48) reduces to

$$p(A) \equiv r(A) \tag{6.49}$$

Thus to evaluate $p(A)$, simply divide $p(\lambda)$ by the characteristic polynomial, and the desired result is obtained by substituting A into the remainder polynomial.

Example 6.9 Continuing with the 2×2 matrix A in Eq. (6.5) having $k(\lambda) = \lambda^2 - 3\lambda - 4$, if

$$p(\lambda) = \lambda^5 + 2\lambda^4 + \lambda^3 - 3\lambda^2 + \lambda - 1$$

then division of $p(\lambda)$ by $k(\lambda)$ produces the remainder $r(\lambda) = 312\lambda + 307$. Hence by (6.49)

$$p(A) = A^5 + 2A^4 + A^3 - 3A^2 + A - I$$
$$= 312A + 307I$$
$$= \begin{bmatrix} 619 & 936 \\ 624 & 931 \end{bmatrix}$$

Problem 6.20 Determine $A^5 - 7A^4 - 18A^3 + 54A^2 + 116A + I_3$ for the matrix in Problem 6.8a.

6.3.4 Companion matrix

We now introduce a matrix with a very simple form which has the useful property that its characteristic polynomial is the same as that of A, so both matrices have the same eigenvalues. For example, when $n = 2$ define

$$C_2 = \begin{bmatrix} 0 & 1 \\ -k_2 & -k_1 \end{bmatrix} \tag{6.50}$$

where the k's are the coefficients in (6.4). It is very easy to see that

$$\det(\lambda I_2 - C_2) = \begin{vmatrix} \lambda & -1 \\ k_2 & \lambda + k_1 \end{vmatrix}$$
$$= \lambda^2 + k_1\lambda + k_2$$

which is the same as $k(\lambda)$ in (6.4) with $n = 2$. Similarly, if

$$C_3 = \begin{bmatrix} 0 & 1 & 0 \\ 0 & 0 & 1 \\ -k_3 & -k_2 & -k_1 \end{bmatrix} \tag{6.51}$$

then

$$\det(\lambda I_3 - C_3) = \lambda^3 + k_1\lambda^2 + k_2\lambda + k_3 \tag{6.52}$$

and the pattern is now emerging. We define C_n to be the $n \times n$ matrix having 1's along the diagonal immediately above the principal diagonal, having last row $[-k_n, -k_{n-1}, \ldots, -k_2, -k_1]$, and zeros everywhere else. Then

$$\det(\lambda I_n - C_n) = \lambda^n + k_1\lambda^{n-1} + \cdots + k_n \equiv k(\lambda) \equiv \det(\lambda I_n - A) \tag{6.53}$$

and for this reason C_n is called the *companion matrix* associated with the polynomial $k(\lambda)$. Because of its simple form C_n is useful in various applications (alternative forms are sometimes used – see Exercise 6.12).

Problem 6.21 Verify that if λ_i is an eigenvalue of C_3 then an associated eigenvector is $[1, \lambda_i, \lambda_i^2]^T$.

Similarly, verify that $m_i = [1, \lambda_i, \lambda_i^2, \ldots, \lambda_i^{n-1}]^T$ is an eigenvector of C_n (notice that m_i is the ith column of the Vandermonde matrix defined in Exercise 4.16).

As an illustration of one of the uses of the companion matrix, we now show that determination of polynomials in C_n is especially easy. For simplicity we consider C_3 in (6.51), but the argument is easily generalized for any value of n. We wish to compute

$$p(C_3) = C_3^2 + p_1 C_3 + p_2 I_3 \tag{6.54}$$

First, if e_i denotes the ith row of I_3 then it is obvious from (6.51) that the first and second rows of C_3 are e_2 and e_3 respectively. Next, if X is *any* 3×3 matrix, then the product $e_i X$ simply gives the ith row of X (see Problem 2.19), so in particular we can write (dropping for convenience the suffix on C):

$$e_1 C = e_2, \qquad e_2 C = e_3 \tag{6.55}$$

Hence the first row of C^2 is given by

$$e_1 C^2 = (e_1 C)C = e_2 C = e_3$$

119

We can now determine the rows ρ_1, ρ_2, ρ_3 of $p(C)$ in (6.54). Since the first rows of C^2, C, and I are e_3, e_2, and e_1 respectively, then clearly

$$\begin{aligned}
\rho_1 &= e_3 + p_1 e_2 + p_2 e_1 \\
&= [0, 0, 1] + p_1[0, 1, 0] + p_2[1, 0, 0] \\
&= [p_2, p_1, 1]
\end{aligned} \tag{6.56}$$

Also,

$$\begin{aligned}
\rho_2 &= e_2 p(C) \\
&= (e_1 C)p(C), \qquad \text{by (6.55)} \\
&= e_1 p(C)C
\end{aligned} \tag{6.57}$$

since $p(C)$ commutes with C (see Problem 6.16). By definition $e_1 p(C) = \rho_1$, so (6.57) reduces to

$$\rho_2 = \rho_1 C \tag{6.58}$$

Similarly,

$$\begin{aligned}
\rho_3 &= e_3 p(C) \\
&= e_2 C p(C), \qquad \text{by (6.55)} \\
&= [e_2 p(C)]C \\
&= \rho_2 C \\
&= \rho_1 C^2, \qquad \text{by (6.58)}
\end{aligned}$$

so we have shown that

$$p(C_3) = \begin{bmatrix} \rho_1 \\ \rho_1 C \\ \rho_1 C^2 \end{bmatrix} \tag{6.59}$$

where ρ_1 is given by (6.56). In general, if C is the $n \times n$ companion matrix, and we form the matrix polynomial

$$p(C) = C^r + p_1 C^{r-1} + p_2 C^{r-2} + \cdots + p_{r-1}C + p_r I \tag{6.60}$$

with $r < n$, then an extension of the argument used to obtain (6.59) shows that

$$p(C) = \begin{bmatrix} \rho_1 \\ \rho_1 C \\ \rho_1 C^2 \\ \vdots \\ \rho_1 C^{n-1} \end{bmatrix} \tag{6.61}$$

where

$$\rho_1 = [p_r, p_{r-1}, \ldots, p_1, 1, 0, \ldots, 0] \tag{6.62}$$

Thus, for any polynomial $p(\lambda)$ having degree less than n, the rows $\rho_1, \rho_2, \ldots, \rho_n$ of the matrix polynomial $p(C)$ can be obtained from (6.62) and the recurrence formula

$$\rho_{i+1} = \rho_i C, \qquad i = 1, 2, \ldots, n - 1 \tag{6.63}$$

and this is much easier in general than constructing $p(A)$ for an arbitrary matrix A.

We can note three interesting points about (6.61). First, it follows by inspection that $p(C) \equiv 0$ only if $\rho_1 = 0$. That is, there is no polynomial of degree less than n such that $p(C) \equiv 0$, so according to the definition in Section 6.3.3., C is nonderogatory. Second, we saw in Section 6.3.3 that the eigenvalues of $p(C)$ are $p(\lambda_1), p(\lambda_2), \ldots, p(\lambda_n)$, where the λ's are the eigenvalues of C, i.e. the roots of the polynomial $k(\lambda)$. Also, (6.21) shows that

$$\det p(C) = p(\lambda_1) p(\lambda_2) \cdots p(\lambda_n)$$

so $p(C)$ is nonsingular if and only if each $p(\lambda_i) \neq 0$. This latter condition means that none of the roots of $k(\lambda)$ can also be a root of $p(\lambda)$, or in other words $k(\lambda)$ and $p(\lambda)$ must have no common factor. Thus $p(C)$ is nonsingular if and only if $k(\lambda)$ and $p(\lambda)$ are relatively prime. In the terminology of Exercise 4.1, $\det p(C)$ is a *resultant* for the two polynomials $k(\lambda)$ and $p(\lambda)$.

The third feature of $p(C)$ which is worth noting is that if we transpose (6.61) we get

$$[p(C)]^{\mathrm{T}} = [\rho_1^{\mathrm{T}}, C^{\mathrm{T}}\rho_1^{\mathrm{T}}, (C^{\mathrm{T}})^2\rho_1^{\mathrm{T}}, \ldots, (C^{\mathrm{T}})^{n-1}\rho_1^{\mathrm{T}}]$$

which has precisely the same form as the controllability matrix introduced in (4.37), with b replaced by ρ_1^{T} and A by C^{T}. This can be developed to produce some interesting results in control theory.

Problem 6.22 Let C be the companion matrix for $k(\lambda) = \lambda^3 + 6\lambda^2 + 11\lambda + 6$. Use (6.62) and (6.63) to construct $p(C)$, where $p(\lambda) = \lambda^2 - \lambda - 2$. Hence determine whether $k(\lambda)$ and $p(\lambda)$ have a common factor.

6.3.5 Kronecker product expressions

If A and B are arbitrary $n \times n$ matrices, then in general there is no way of determining the eigenvalues either of the product AB or of the sum $A + B$ in terms of the eigenvalues of A and B. However, as we have seen in other situations, these difficulties can be avoided if we use the Kronecker product.

Let A be $n \times n$ with eigenvalues and eigenvectors λ_i, u_i ($i = 1, 2, \ldots, n$) and let B be $m \times m$ with eigenvalues and eigenvectors μ_j, y_j ($j = 1, 2, \ldots, m$), so by definition $Au_i = \lambda_i u_i$, $By_j = \mu_j y_j$. Hence, using

(2.75) we have

$$(A \otimes B)(u_i \otimes y_j) = (Au_i) \otimes (By_j)$$
$$= (\lambda_i u_i) \otimes (\mu_j y_j)$$
$$= (\lambda_i \mu_j)(u_i \otimes y_j) \tag{6.64}$$

showing that the mn eigenvalues and eigenvectors of $A \otimes B$ are simply $\lambda_i \mu_j$ and $u_i \otimes y_j$, for $i = 1, \ldots, n$ and $j = 1, \ldots, m$.

A similar result can be obtained for a Kronecker 'sum' involving A and B. We use the matrix

$$D = A \otimes I_m + I_n \otimes B^{\mathrm{T}} \tag{6.65}$$

which arose in (5.52) in connection with a linear matrix equation. We shall show that the eigenvalues of the $mn \times mn$ matrix D are all possible sums $\lambda_i + \mu_j$. (The occurrence of B^{T} instead of B in (6.65) is of no consequence – we saw at the beginning of Section 6.3.3 that the eigenvalues of B^{T} are the same as those of B.)

The method of proof is quite ingenious. Let ϵ be an arbitrary scalar parameter, and consider the product

$$(I_n + \epsilon A) \otimes (I_m + \epsilon B^{\mathrm{T}}) = I_n \otimes I_m + \epsilon(A \otimes I_m + I_n \otimes B^{\mathrm{T}}) + \epsilon^2 A \otimes B^{\mathrm{T}}$$
$$= I_n \otimes I_m + \epsilon D + \epsilon^2 A \otimes B^{\mathrm{T}} \tag{6.66}$$

By the result of Problem 6.15 the eigenvalues of $I + \epsilon A$ and $I + \epsilon B^{\mathrm{T}}$ are $1 + \epsilon \lambda_i$ and $1 + \epsilon \mu_j$ respectively. By the foregoing argument, expressed in (6.64), the eigenvalues of the matrix on the left-hand side of (6.66) are

$$(1 + \epsilon \lambda_i)(1 + \epsilon \mu_j) = 1 + \epsilon(\lambda_i + \mu_j) + \epsilon^2 \lambda_i \mu_j \tag{6.67}$$

Since ϵ is *arbitrary*, it follows by comparing terms in ϵ on the right-hand sides of (6.66) and (6.67) that D has eigenvalues $\lambda_i + \mu_j$.

Since $\det D$ is equal to the product of the eigenvalues of D (see (6.21)) it follows that D is nonsingular if and only if $\lambda_i + \mu_j \neq 0$, for all i and j, and this is the condition for the solution X of the matrix equation (5.51) (i.e., $AX + XB = C$) to be unique.

Problem 6.23 Write down in terms of A and B a matrix whose eigenvalues are the mn numbers $\lambda_i^2 + \mu_j^2$, $i = 1, \ldots, n$; $j = 1, \ldots, m$.

Problem 6.24 If E is a $p \times p$ matrix having eigenvalues and eigenvectors ν_k, z_k, what are the eigenvalues and eigenvectors of $A \otimes B \otimes E$?

6.4 Similarity

We shall assume *throughout* this section that all the eigenvalues λ_i of A are distinct (i.e., $\lambda_i \neq \lambda_j$ for all $i \neq j$). This is often the case in practical problems.

6.4.1 Definition

Two $n \times n$ matrices A and B are called *similar* if

$$B = P^{-1}AP \qquad (6.68)$$

where P is an arbitrary nonsingular $n \times n$ matrix. The relationship between A and B is called that of *similarity*. The characteristic equation of B is

$$0 = |\lambda I_n - P^{-1}AP| \qquad (6.69)$$

and we now do some juggling with (6.69) to prove that the eigenvalues of B are the same as those of A. First, replace I_n in (6.69) by $P^{-1}P$ and factorize inside the determinant sign:

$$0 = |P^{-1}(\lambda I - A)P| \qquad (6.70)$$

Then use the result (4.33) on the determinant of a product, thereby reducing (6.70) to

$$0 = |P^{-1}||\lambda I - A||P|$$
$$= |\lambda I - A|$$

since $|P^{-1}| = 1/|P|$ (see (4.42)). This shows that the characteristic equations of B and A are identical, as required. Furthermore, if u_i are the eigenvectors of A then

$$B(P^{-1}u_i) = P^{-1}AP(P^{-1}u_i)$$
$$= P^{-1}Au_i$$
$$= P^{-1}\lambda_i u_i = \lambda_i(P^{-1}u_i)$$

showing that the eigenvectors of B are $P^{-1}u_i$, for $i = 1, 2, \ldots, n$.

The converse result also holds, namely, that if B is a matrix having the same eigenvalues as A then Eq. (6.68) holds for some P. (It is worth remarking that this does not apply in general if any of the λ's are repeated, but we shan't discuss this further.)

Problem 6.25 Prove that if A and B are similar then (a) $\det A = \det B$, (b) $\mathrm{tr}(A) = \mathrm{tr}(B)$.

Problem 6.26 Prove that if all the n eigenvalues of A are distinct then there exists an $n \times n$ nonsingular matrix X such that $A^T X - XA = 0$.

Problem 6.27 If A and B are similar, determine which of the following pairs is similar: (a) A^T and B^T; (b) A^k and B^k ($k = $ positive integer); (c) A^{-1} and B^{-1} (assuming A is nonsingular).

123

6.4.2 Diagonalization

We now show how to choose P in (6.68) so that B is a diagonal matrix. The development depends on the fact (stated here without proof), that provided the λ's are distinct then the eigenvectors $u_1, \ldots, u_n$ of A are linearly independent. From the discussion in Section 5.7 this means that the matrix formed by putting the u's side by side:

$$T = [u_1, u_2, \ldots, u_n] \tag{6.71}$$

has rank n, i.e., is nonsingular. It is T which we use for P in (6.68), which becomes

$$
\begin{aligned}
T^{-1}AT &= T^{-1}A[u_1, u_2, \ldots, u_n] \\
&= T^{-1}[Au_1, Au_2, \ldots, Au_n] \\
&= T^{-1}[\lambda_1 u_1, \lambda_2 u_2, \ldots, \lambda_n u_n], \text{ by (6.32)} \\
&= [\lambda_1 T^{-1}u_1, \lambda_2 T^{-1}u_2, \ldots, \lambda_n T^{-1}u_n]
\end{aligned}
\tag{6.72}
$$

However, we also have

$$
\begin{aligned}
I_n = T^{-1}T &= T^{-1}[u_1, u_2, \ldots, u_n] \\
&= [T^{-1}u_1, T^{-1}u_2, \ldots, T^{-1}u_n]
\end{aligned}
\tag{6.73}
$$

and comparison of (6.72) and (6.73) shows that the ith column in (6.72) is simply λ_i times the ith column in (6.73), and this latter is the ith column of I_n. Therefore (6.72) reduces to

$$T^{-1}AT = \text{diag}[\lambda_1, \lambda_2, \ldots, \lambda_n] \equiv \Lambda, \tag{6.74}$$

showing that when all the λ's are distinct then A is similar to Λ, the diagonal matrix of its eigenvalues. The similarity transformation (6.74) of A into Λ is called *diagonalization*, and has many important applications.

Example 6.10 Return to the matrix A in Eq. (6.5) having $\lambda_1 = -1$, $\lambda_2 = 4$, and use the eigenvectors obtained in (6.7):

$$u_1 = t_1 \begin{bmatrix} 1 \\ -\frac{2}{3} \end{bmatrix}, \qquad u_2 = t_2 \begin{bmatrix} 1 \\ 1 \end{bmatrix}$$

For convenience set $t_1 = 3$, $t_2 = 1$, so from (6.71)

$$T = \begin{bmatrix} 3 & 1 \\ -2 & 1 \end{bmatrix}, \qquad T^{-1} = \frac{1}{5}\begin{bmatrix} 1 & -1 \\ 2 & 3 \end{bmatrix} \tag{6.75}$$

the inverse being obtained using (4.7). It is easily verified that $T^{-1}AT = \text{diag}[-1, 4]$.

Notice that any suitable vectors u_i in (6.71) will do – there is no need to use normalized eigenvectors.

It should be noted that if some of the λ's are repeated then A is similar to the diagonal matrix Λ only if A possesses n linearly independent eigenvectors (see Exercise 6.28).

Problem 6.28 Determine eigenvectors for the matrix A in Problem 6.8b. Calculate T^{-1}, and hence verify (6.74) in this case.

Problem 6.29 Use (6.74) to show that $A = T\Lambda T^{-1}$, and hence prove that

$$A^k = T\Lambda^k T^{-1} \tag{6.76}$$

for any positive integer k. Use (6.76) to calculate A^3 for the matrix A in Problem 6.28. Similarly, show that for the matrix polynomial in (6.35), $p(A) = Tp(\Lambda)T^{-1}$.

Problem 6.30 With the notation of (6.74), if A is nonsingular, define

$$X = T\{\text{diag}\,[\pm\lambda_1^{1/2}, \pm\lambda_2^{1/2}, \ldots, \pm\lambda_n^{1/2}]\}T^{-1} \tag{6.77}$$

and show that $X^2 = A$. Thus X can be thought of as a 'square root' of A.

It is worth mentioning here that for T in (6.71) (the columns of which are right eigenvectors of A) then the rows $v_1, \ldots, v_n$ of its inverse T^{-1} are *left* eigenvectors of A (defined in Eq. (6.9)). To show this, rearrange (6.74) as $T^{-1}A = \Lambda T^{-1}$ and compare the ith rows on each side to obtain $v_i A = \lambda_i v_i$, as required. As an illustration, the rows of T^{-1} in Example 6.10 are

$$v_1 = \frac{1}{5}[1, -1], \qquad v_2 = \frac{1}{5}[2, 3]$$

and using A in (6.5) it is easy to verify that

$$v_1 A = (-1)v_1, \qquad v_2 A = 4v_2$$

showing that v_1, v_2 are left eigenvectors. For a converse result, see Exercise 6.15.

Problem 6.31 Verify that in Problem 6.28 the rows of T^{-1} are left eigenvectors of A.

We close this section by pointing out that similarity in (6.68) is a special case of equivalence of matrices, defined in (5.14), since in (6.68) both P^{-1} and P are nonsingular. Thus all the properties of equivalent matrices also apply to similar matrices. For example, two similar matrices have the same rank.

6.4.3 Hermitian and symmetric matrices

When A is hermitian (or real symmetric) the diagonalization formula (6.74) can be simplified. To develop this, let

$$Au_i = \lambda_i u_i, \qquad Au_j = \lambda_j u_j, \qquad i \neq j \tag{6.78}$$

We saw in Section 6.3.2 that all the λ's are real. Premultiply the first equation in (6.78) by u_j^*:

$$u_j^* A u_i = \lambda_i u_j^* u_i \tag{6.79}$$

Take the conjugate transpose of the second equation in (6.78) and then postmultiply by u_i to get

$$u_j^* A u_i = \lambda_j u_j^* u_i \tag{6.80}$$

(using $A^* = A, \bar{\lambda}_j = \lambda_j$). Comparison of (6.79) and (6.80) reveals that

$$\lambda_i u_j^* u_i = \lambda_j u_j^* u_i \tag{6.81}$$

and since $\lambda_i \neq \lambda_j$ (by assumption) we conclude from (6.81) that

$$u_j^* u_i = 0, \qquad \text{for all } i \neq j \tag{6.82}$$

Any two vectors whose scalar product is zero are called *mutually orthogonal*, by analogy with the geometrical result in two and three dimensions. Thus (6.82) shows that all pairs of eigenvectors of any hermitian matrix (having distinct eigenvalues) are mutually orthogonal. If the u's are normalized, i.e., by (6.8)

$$u_i^* u_i = 1, \qquad i = 1, \ldots, n \tag{6.83}$$

then for T in (6.71) we have

$$T^*T = \begin{bmatrix} u_1^* \\ \vdots \\ u_n^* \end{bmatrix} [u_1, \ldots, u_n] = I_n$$

using (6.82) and (6.83). This shows that T in this case is a unitary matrix (see Exercise 4.11), and $T^{-1} = T^*$. Hence the diagonalization (6.74) becomes

$$T^*AT = \Lambda \tag{6.84}$$

and A is said to be *unitarily* similar to Λ. When A is real symmetric all the u's are real vectors, and T is orthogonal (i.e., $T^{-1} = T^T$) so that (6.84) is replaced by *orthogonal* similarity

$$T^T A T = \Lambda \tag{6.85}$$

The transformations (6.84) and (6.85) are simpler to apply than the general case (6.74) since there is no need to calculate T^{-1}.

Example 6.11 For the real symmetric matrix

$$A = \begin{bmatrix} 1 & -2 \\ -2 & -2 \end{bmatrix} \tag{6.86}$$

the usual calculations give $\lambda_1 = 2$, $\lambda_2 = -3$, with corresponding normalized eigenvectors

$$u_1 = \frac{1}{\sqrt{5}}\begin{bmatrix} 2 \\ -1 \end{bmatrix}, \qquad u_2 = \frac{1}{\sqrt{5}}\begin{bmatrix} 1 \\ 2 \end{bmatrix}, \qquad T = [u_1, u_2]$$

Hence in this case (6.85) is

$$\underset{T^T}{\frac{1}{\sqrt{5}}\begin{bmatrix} 2 & -1 \\ 1 & 2 \end{bmatrix}} \underset{A}{\begin{bmatrix} 1 & -2 \\ -2 & -2 \end{bmatrix}} \underset{T}{\frac{1}{\sqrt{5}}\begin{bmatrix} 2 & 1 \\ -1 & 2 \end{bmatrix}} = \underset{\Lambda}{\begin{bmatrix} 2 & 0 \\ 0 & -3 \end{bmatrix}}$$

Problem 6.32 Show that the matrix

$$A = \begin{bmatrix} 11 & 2 & 8 \\ 2 & 2 & -10 \\ 8 & -10 & 5 \end{bmatrix} \tag{6.87}$$

has one eigenvalue equal to -9, and calculate the other eigenvalues, and normalized eigenvectors. Hence carry out the diagonalization (6.85).

It is important to note that it can be shown that *every* hermitian (or real symmetric) matrix can be reduced by unitary (or orthogonal) similarity to a diagonal matrix, even if some of the λ's occur more than once (in fact, unitary similarity also holds for *normal* matrices, defined in Exercise 2.10). As remarked at the end of the preceding section, because A is similar to Λ we have $R(A) = R(\Lambda)$, and since Λ is diagonal its rank is equal to the number of nonzero diagonal elements. Thus, for *any* hermitian (or real symmetric) matrix, its rank is equal to the total number of its nonzero eigenvalues, including repetitions.

Problem 6.33 Let A be an $n \times n$ matrix having all its elements equal to unity. By considering $R(A)$ deduce that A has only one nonzero eigenvalue. Determine its value by applying (6.23).

6.4.4 Transformation to companion form

If A has characteristic polynomial $k(\lambda)$ in (6.53), then the companion matrix C defined in Section 6.3.4 has the same eigenvalues as A, so if all these values are distinct then A and C are similar. In fact the eigenvectors of C were found in Problem 6.21 to be the columns m_i of the Vandermonde matrix

$$V_n = [m_1, m_2, \ldots, m_n] \tag{6.88}$$

where

$$m_i = [1, \lambda_i, \lambda_i^2, \ldots, \lambda_i^{n-1}]^\mathrm{T}$$

From (4.81) V_n is nonsingular provided all the λ's are distinct, so in this case the similarity transformation between C and Λ is

$$V_n^{-1}CV_n = \Lambda \tag{6.89}$$

Combining (6.89) with (6.74) gives

$$C = V_n\Lambda V_n^{-1} = V_n T^{-1}ATV_n^{-1}$$
$$= (TV_n^{-1})^{-1}A(TV_n^{-1}) \tag{6.90}$$

In (6.90) T is the matrix (6.71) of eigenvectors of A, and V_n is defined by (6.88) and (6.89), so the similarity transformation from A to C can be determined.

Example 6.12 For A in Eq. (6.5) we found $\lambda_1 = -1$, $\lambda_2 = 4$ so

$$V_2 = \begin{bmatrix} 1 & 1 \\ -1 & 4 \end{bmatrix}, \qquad V_2^{-1} = \frac{1}{5}\begin{bmatrix} 4 & -1 \\ 1 & 1 \end{bmatrix}$$

the inverse being obtained from (4.7). The matrix T is given in (6.75) so

$$TV_2^{-1} = \frac{1}{5}\begin{bmatrix} 13 & -2 \\ -7 & 3 \end{bmatrix}$$

and substitution into (6.90) gives

$$(TV_2^{-1})^{-1}A(TV_2^{-1}) = \begin{bmatrix} 0 & 1 \\ 4 & 3 \end{bmatrix}$$

which is the companion matrix associated with the characteristic polynomial $\lambda^2 - 3\lambda - 4$ of A.

Problem 6.34 Carry out the transformation (6.90) for A in Problem 6.1.

When some of the λ's are repeated, the above method is no longer valid since V_n is singular. In general, A is similar to the companion matrix C associated with its characteristic polynomial if and only if A is nonderogatory (in Exercise 6.21, take B to be this matrix C).

6.5 Solution of linear differential and difference equations

An important use of (6.74) is in solving a set of linear differential equations

$$\dot{x} = Ax \tag{6.91}$$

where A is a constant $n \times n$ matrix and $x = [x_1(t), \ldots, x_n(t)]^T$. Two examples of such sets of equations were given in Section 6.2. Let

$$y = T^{-1}x, \quad x = Ty \tag{6.92}$$

where T is the matrix of eigenvectors in (6.71). Then

$$\dot{y} = T^{-1}\dot{x} = T^{-1}Ax$$
$$= T^{-1}ATy$$
$$= \Lambda y \tag{6.93}$$

the last step following by (6.74). Since Λ is diagonal the equations represented by (6.93) have the very simple form

$$\dot{y}_i = \lambda_i y_i, \quad i = 1, 2, \ldots, n \tag{6.94}$$

As in (6.15) in Section 6.2, the solution of (6.94) is

$$y_i = \alpha_i \, e^{\lambda_i t}, \quad i = 1, 2, \ldots, n \tag{6.95}$$

where the α's are arbitrary constants of integration. Hence by (6.92) the general solution of (6.91) is

$$x(t) = T[\alpha_1 e^{\lambda_1 t}, \alpha_2 e^{\lambda_2 t}, \ldots, \alpha_n e^{\lambda_n t}]^T \tag{6.96}$$

Example 6.13 If A is the matrix in Eq. (6.5), then the Eqs (6.91) are

$$\frac{dx_1}{dt} = x_1 + 3x_2, \quad \frac{dx_2}{dt} = 2x_1 + 2x_2 \tag{6.97}$$

In this case $\lambda_1 = -1$, $\lambda_2 = 4$ and T is given in (6.75), so by (6.96) the general solution of (6.97) is

$$\begin{bmatrix} x_1(t) \\ x_2(t) \end{bmatrix} = \begin{bmatrix} 3 & 1 \\ -2 & 1 \end{bmatrix} \begin{bmatrix} \alpha_1 e^{-t} \\ \alpha_2 e^{4t} \end{bmatrix}$$

or

$$x_1 = 3\alpha_1 e^{-t} + \alpha_2 e^{4t}, \quad x_2 = -2\alpha_1 e^{-t} + \alpha_2 e^{4t} \tag{6.98}$$

where α_1 and α_2 are arbitrary constants whose values can be determined from given initial conditions $x_1(0)$ and $x_2(0)$.

Problem 6.35 Verify by differentiation that (6.98) is indeed the solution of (6.97).

Problem 6.36 Solve (6.91) when A is the matrix used in Problem 6.8b (use the results obtained in Problem 6.28).

A very similar development holds for linear difference equations of the form

$$X(k+1) = AX(k), \qquad k = 0, 1, 2, \dots \tag{6.99}$$

where $X(k) = [X_1(k), X_2(k), \dots, X_n(k)]^T$. For examples of applications of (6.99), see Exercises 2.11 and 6.10. Using the transformation $Y(k) = T^{-1}X(k)$, $X(k) = TY(k)$, leads to

$$Y(k+1) = \Lambda Y(k)$$

or in component form

$$Y_i(k+1) = \lambda_i Y_i(k), \qquad i = 1, 2, \dots, n \tag{6.100}$$

It is easily verified that Eqs (6.100) have the general solution

$$Y_i(k) = \beta_i \lambda_i^k, \qquad i = 1, 2, \dots, n$$

where the β's are arbitrary constants, so the general solution of (6.99) is

$$X(k) = T[\beta_1 \lambda_1^k, \beta_2 \lambda_2^k, \dots, \beta_n \lambda_n^k]^T \tag{6.101}$$

Example 6.14 Consider the difference equations

$$X_1(k+1) = X_1(k) + 3X_2(k), \qquad X_2(k+1) = 2X_1(k) + 2X_2(k)$$

Once again A is the matrix in (6.5) with $\lambda_1 = -1$, $\lambda_2 = 4$ and T is given by (6.75). Thus from (6.101) the general solution of this pair of difference equations is

$$\begin{bmatrix} X_1(k) \\ X_2(k) \end{bmatrix} = \begin{bmatrix} 3 & 1 \\ -2 & 1 \end{bmatrix} \begin{bmatrix} \beta_1(-1)^k \\ \beta_2(4)^k \end{bmatrix}$$

Problem 6.37 Verify by substitution that the solution obtained in the preceding example is indeed correct.

Problem 6.38 Determine the eigenvalues of the matrix A in (2.74) (notice that A is in companion form). Apply (6.101) to obtain the general expression for the solution of (2.74). Using the conditions $X_1(1) = 1$, $X_1(2) = 1$ determine β_1 and β_2, and hence show that the kth Fibonacci number is

$$x_k = X_1(k) = \frac{1}{\sqrt{5}}\left[\left(\frac{1+\sqrt{5}}{2}\right)^k - \left(\frac{1-\sqrt{5}}{2}\right)^k\right]$$

$$= \frac{1}{2^{k-1}}\left[k + \binom{k}{3}5 + \binom{k}{5}5^2 + \binom{k}{7}5^3 + \cdots\right]$$

where $\binom{k}{r}$ denotes the usual binomial coefficient.

130

6.6 Calculation of eigenvalues and eigenvectors

As remarked previously, it is not recommended that the eigenvalues of a matrix be calculated by solving the characteristic equation. This is because of difficulties both in computing the coefficients in the characteristic polynomial, and in applying root-finding methods (these may also be very sensitive to errors in the coefficients). Instead, some iterative process is applied to the matrix itself. We confine ourselves to describing one simple procedure which can easily be carried out with a pocket calculator. Details of computer methods used in real-life problems can be found in numerical textbooks such as Goult *et al.* (1974) and Stewart (1973).

6.6.1 Power method

Suppose that all the eigenvalues of a real matrix A are distinct, and that the eigenvalue having largest modulus is real. Let this value (termed the *dominant* eigenvalue) be denoted by λ_1, so that the remaining eigenvalues $\lambda_2, \ldots, \lambda_n$ satisfy

$$|\lambda_i| < |\lambda_1|, \qquad i = 2, 3, \ldots, n \tag{6.102}$$

Since corresponding eigenvectors $u_1, \ldots, u_n$ are linearly independent they form a basis, so any arbitrary column n-vector X_0 can be expressed as a linear combination

$$X_0 = \alpha_1 u_1 + \alpha_2 u_2 + \cdots + \alpha_n u_n \tag{6.103}$$

If $X_1 = AX_0$, then since $Au_i = \lambda_i u_i$ we have

$$X_1 = \alpha_1 A u_1 + \alpha_2 A u_2 + \cdots + \alpha_n A u_n$$
$$= \alpha_1 \lambda_1 u_1 + \alpha_2 \lambda_2 u_2 + \cdots + \alpha_n \lambda_n u_n$$

Similarly if $X_2 = AX_1$, $X_3 = AX_2$, and so on, we obtain

$$X_k = \alpha_1 \lambda_1^k u_1 + \alpha_2 \lambda_2^k u_2 + \cdots + \alpha_n \lambda_n^k u_n, \quad k = 1, 2, 3, \ldots \tag{6.104}$$

and hence

$$\frac{1}{\lambda_1^k} X_k = \alpha_1 u_1 + \alpha_2 \left(\frac{\lambda_2}{\lambda_1}\right)^k u_1 + \cdots + \alpha_n \left(\frac{\lambda_n}{\lambda_1}\right)^k u_n \tag{6.105}$$

Since from (6.102) $|\lambda_i/\lambda_1| < 1$ for each $i > 1$, it follows that as $k \to \infty$ the terms $(\lambda_i/\lambda_1)^k$ tend to zero, so (6.105) gives

$$\frac{1}{\lambda_1^k} X_k \to \alpha_1 u_1, \quad \text{as } k \to \infty \tag{6.106}$$

Thus provided $\alpha_1 \neq 0$, (6.106) shows that X_k tends to a multiple of the eigenvector u_1 as $k \to \infty$. However, as k increases the elements of X_k can become large, so to avoid this it is preferable to scale X_k at each step so that its largest element is unity. We therefore define a modified sequence $X_1, X_2, X_3, \ldots$, as follows:

$$\left.\begin{aligned} Y_{k+1} &= AX_k, \qquad k = 0, 1, 2, \ldots \\ \beta_{k+1} &= \text{element of } Y_{k+1} \text{ having} \\ & \qquad \text{largest modulus} \\ X_{k+1} &= \frac{1}{\beta_{k+1}} Y_{k+1} \end{aligned}\right\} \qquad (6.107)$$

As $k \to \infty$ we still have X_k tending to some multiple of u_1, so we can write $X_k \to pu_1$ and then (6.107) implies

$$Y_k \to A(pu_1) = p(Au_1) = p(\lambda_1 u_1) = \lambda_1(pu_1)$$

showing that $Y_k \to \lambda_1 X_k$ as $k \to \infty$. Because the largest element of X_k is unity, it follows that the element of Y_k having largest modulus approaches λ_1, i.e., $\beta_k \to \lambda_1$ as $k \to \infty$. Thus β_k and X_k defined by (6.107) provide successively better approximations to λ_1 and u_1, and the process is terminated when X_{k+1} and X_k are sufficiently close.

Example 6.15 For the matrix A in (6.5) start by choosing an arbitrary vector $X_0 = [1, \frac{1}{2}]^T$. Then (6.107) gives

$$Y_1 = AX_0 = \begin{bmatrix} 2.5 \\ 3 \end{bmatrix}, \qquad \beta_1 = 3, \qquad X_1 = \frac{1}{3} Y_1 = \begin{bmatrix} 0.833 \\ 1 \end{bmatrix}$$

$$Y_2 = AX_1 = \begin{bmatrix} 3.833 \\ 3.667 \end{bmatrix}, \qquad \beta_2 = 3.833, \qquad X_2 = \begin{bmatrix} 1 \\ 0.957 \end{bmatrix}$$

$$Y_3 = AX_2 = \begin{bmatrix} 3.871 \\ 3.914 \end{bmatrix}, \qquad \beta_3 = 3.914, \qquad X_3 = \begin{bmatrix} 0.989 \\ 1 \end{bmatrix}$$

and so on. Working to three decimal places, continuation of this process gives $X_6 = [1, 1]^T$ and $\beta_7 = 4$, which agree with the exact values found in Example 6.1.

Clearly in general the speed of convergence will depend on how quickly the terms in (6.105) tend to zero, that is, upon the magnitudes of the ratios $|\lambda_i/\lambda_1|$ for $i > 1$. The method breaks down if by chance α_1 in (6.103) is zero, but in any case this can be overcome simply by choosing a different initial vector and starting again.

Once λ_1 is known an $(n-1) \times (n-1)$ matrix A' can be constructed having eigenvalues $\lambda_2, \lambda_3, \ldots, \lambda_n$ (see Exercise 6.27). If λ_2 is real and $|\lambda_2| > |\lambda_i|$, for all $i > 2$, then the power method can be used to estimate λ_2 and u_2, and so on.

Problem 6.39 Use the power method with $X_0 = [1, 1]^T$ to estimate the dominant eigenvalue and corresponding eigenvector for $A = \begin{bmatrix} 4 & 1 \\ 2 & 5 \end{bmatrix}$.

Problem 6.40 Let the eigenvalue λ_n of A having smallest modulus be real. Prove that the dominant eigenvalue of A^{-1} is $1/\lambda_n$ (hint: see Problem 6.13).

Hence compute an estimate of the smaller eigenvalue of A in Problem 6.39 by applying the power method to A^{-1}, with $X_0 = [1, 0]^T$.

6.6.2 Other methods

A great deal of effort has been devoted to devising effective computer algorithms for calculating eigenvalues and eigenvectors. One important class of methods is based on the use of similarity transformations of the form $A_{k+1} = Q_k^T A_k Q_k$, $k = 0, 1, 2, \ldots$, with $A_0 = A$ and Q_k orthogonal. The matrices Q_k are chosen so that A_k approaches a simple form (e.g., diagonal) whose eigenvalues are readily obtainable. For example, in the so-called QR-type methods, which are widely used, a sequence is generated by

$$A_k = Q_k R_k, \qquad R_k Q_k = A_{k+1}, \qquad k = 0, 1, 2, \ldots \qquad (6.108)$$

with Q_k orthogonal and R_k upper triangular, and then A_k tends to upper triangular form as $k \to \infty$. These transformation methods are always convergent, and produce a full set of eigenvalues and eigenvectors.

As in the case of linear equations, any large practical problem will require the use of an automatic computer and a proven library program.

Problem 6.41 Show that in (6.108) A_{k+1} is similar to A_k.

6.7 Iterative solution of linear equations

We are now able to consider some other methods for solving the n linear equations in n unknowns in our usual form

$$Ax = b \qquad (6.109)$$

The procedures we shall describe are different in nature from those considered in Chapter 3, where the solution was obtained after a *fixed* finite number of operations (such methods are termed 'direct'). Instead, with an iterative approach, an initial guess $x^{(0)}$ for the solution of (6.109) is used to obtain a better approximation $x^{(1)}$, which in turn is used to produce $x^{(2)}$, and so on. The aim is that the sequence of vectors $\{x^{(k)}\}$, $k = 1, 2, 3, \ldots$ converges to the exact solution x of (6.109), by which we mean that each element of $x^{(k)}$ converges to the corresponding element of the exact solution x as $k \to \infty$.

6.7.1 Gauss–Seidel and Jacobi methods

The first step is to split up A into two parts

$$A = B - C \tag{6.110}$$

where B is chosen to be nonsingular and to have an appropriate form which will be described later. Substitution of (6.110) into (6.109) gives

$$Bx = b + Cx \tag{6.111}$$

and the iterative procedure is defined by

$$Bx^{(k+1)} = Cx^{(k)} + b, \qquad k = 0, 1, 2, \ldots \tag{6.112}$$

It is, of course, necessary to determine conditions under which $x^{(k)}$ in (6.112) tends to x as $k \to \infty$, and this problem provides an interesting application of diagonalization. Subtract (6.111) from (6.112) to obtain

$$B(x^{(k+1)} - x) = C(x^{(k)} - x)$$

or

$$x^{(k+1)} - x = L(x^{(k)} - x) \tag{6.113}$$

where $L = B^{-1}C$. Substituting $k = 0, 1, 2, \ldots$ in (6.113) gives in turn

$$x^{(1)} - x = L(x^{(0)} - x)$$
$$x^{(2)} - x = L(x^{(1)} - x) = L^2(x^{(0)} - x)$$
$$x^{(3)} - x = L(x^{(2)} - x) = L^3(x^{(0)} - x)$$
$$\vdots$$
$$x^{(k)} - x = L^k(x^{(0)} - x) \tag{6.114}$$

We require $x^{(k)} - x$ to tend to zero as $k \to \infty$. For arbitrary $x^{(0)}$ ($\neq x$), (6.114) therefore shows that we must have $L^k \to 0$ as $k \to \infty$ (i.e., every element of L^k must tend to zero as $k \to \infty$). Now suppose the eigenvalues of L are $l_1, l_2, \ldots, l_n$ and are distinct. Then by the diagonalization formula (6.74)

$$L = T[\text{diag}(l_1, l_2, \ldots, l_n)]T^{-1}$$

so as in (6.76)

$$L^k = T[\text{diag}(l_1^k, l_2^k, \ldots, l_n^k)]T^{-1} \tag{6.115}$$

Since T in (6.115) is a constant matrix it follows that for $L^k \to 0$ we must have each term $l_i^k \to 0$, and the condition for this is that $|l_i| < 1$ for each i. Thus we have shown that provided all the eigenvalues of L have modulus less than unity, then the process defined in (6.112) converges to the solution of (6.109) (the argument can be extended to cover the case when some l's are repeated).

Unfortunately this test for convergence is of limited value, since calculation of the l's may involve more effort than solving the original equations (6.109)! A more useful criterion, which, however, only gives a *sufficient* condition for convergence of (6.114), is that

$$\|L\| < 1 \tag{6.116}$$

where $\|L\|$, the *euclidean norm* of the matrix L, is defined in a similar way to the norm of a vector in (6.8):

$$\|L\| = \left(\sum_{i=1}^{n} \sum_{j=1}^{n} |l_{ij}|^2 \right)^{1/2} \tag{6.117}$$

Notice (see Exercise 2.7c) that $\|L\| = [\mathrm{tr}(AA^*)]^{1/2}$. Although (6.116) is too restrictive it has the advantage of being very easy to apply.

We now give two ways of choosing B in (6.110). Let $A = [a_{ij}]$, and assume all $a_{ii} \neq 0$. Define

$$D = \mathrm{diag}[a_{11}, a_{22}, \ldots, a_{nn}] \tag{6.118}$$

and a lower triangular matrix $E = [e_{ij}]$ with all $e_{ii} = 0$ and $e_{ij} = -a_{ij}$, $j < i$, and an upper triangular matrix $F = [f_{ij}]$ with all $f_{ii} = 0$ and $f_{ij} = -a_{ij}$, $j > i$. In other words, E is the 'lower part' of $-A$, F is the 'upper part' of $-A$, and D is the 'diagonal part' of A, so that

$$A = D - E - F \tag{6.119}$$

For example, if

$$A = \begin{bmatrix} 4 & 1 & -2 \\ -1 & 6 & -1 \\ 1 & -3 & 8 \end{bmatrix} \tag{6.120}$$

then

$$D = \mathrm{diag}[4, 6, 8] \tag{6.121}$$

$$E = \begin{bmatrix} 0 & 0 & 0 \\ 1 & 0 & 0 \\ -1 & 3 & 0 \end{bmatrix}, \quad F = \begin{bmatrix} 0 & -1 & 2 \\ 0 & 0 & 1 \\ 0 & 0 & 0 \end{bmatrix} \tag{6.122}$$

If in (6.110) we take

$$B = D - E, \quad C = F \tag{6.123}$$

the scheme (6.112) is called the *Gauss–Seidel* method. If we take

$$B = D, \quad C = E + F \tag{6.124}$$

the scheme (6.112) is called *Jacobi's* method. The reason for these choices is that in both cases B has a simple form (respectively triangular

135

and diagonal) so that the equations (6.112) are easily solved at each iteration. In fact some simple explicit formulae can be derived as follows. For Jacobi's method (6.112) and (6.124) give

$$x^{(k+1)} = B^{-1}(Cx^{(k)} + b)$$

$$= D^{-1}[(E + F)x^{(k)} + b] \qquad (6.125)$$

and since D is diagonal we have $D^{-1} = \text{diag}[1/a_{11}, \ldots, 1/a_{nn}]$. Also, $E + F$ is simply the matrix $(-A)$ with all the elements on the principal diagonal replaced by zeros. Thus when (6.125) is written out in component form we obtain

$$x_i^{(k+1)} = \left(-\sum_{\substack{j=1 \\ j \neq i}}^{n} a_{ij}x_j^{(k)} + b_i\right) \Big/ a_{ii}, \qquad i = 1, 2, \ldots, n \qquad (6.126)$$

It is illuminating to compare (6.126) with the ith equation in (6.109), which is

$$\sum_{j=1}^{n} a_{ij}x_j = b_i$$

or, on rearranging to express x_i in terms of the other x's (assuming $a_{ii} \neq 0$):

$$x_i = \left(-\sum_{\substack{j=1 \\ j \neq i}}^{n} a_{ij}x_j + b_i\right) \Big/ a_{ii} \qquad (6.127)$$

Clearly (6.126) is the same as (6.127), except for the superscripts. This shows that the Jacobi process computes the value of x_i at the $(k+1)$th step by simply substituting into the ith equation the values of the x's obtained at the *previous* kth stage.

For the Gauss–Seidel method, (6.112) and (6.123) give

$$(D - E)x^{(k+1)} = Fx^{(k)} + b$$

or

$$x^{(k+1)} = D^{-1}[Ex^{(k+1)} + Fx^{(k)} + b]$$

which in component form becomes

$$x_i^{(k+1)} = \left(-\sum_{j=1}^{i-1} a_{ij}x_j^{(k+1)} - \sum_{j=i+1}^{n} a_{ij}x_j^{(k)} + b_i\right) \Big/ a_{ii} \qquad (6.128)$$

Again, a comparison with (6.127) shows that $x_i^{(k+1)}$ is obtained by substituting previously calculated values into the ith equation, but this time $x_i^{(k+1)}$ is evaluated using the values $x_1^{(k+1)}, x_2^{(k+1)}, \ldots, x_{i-1}^{(k+1)}$ as soon as they are obtained. The following example will serve to clarify this. (For both methods it is convenient to start with $x^{(0)} = 0$.)

EIGENVALUES AND EIGENVECTORS

Example 6.16 Consider the equations

$$4x_1 + x_2 - 2x_3 = 4$$
$$-x_1 + 6x_2 - x_3 = -9.5$$
$$x_1 - 3x_2 + 8x_3 = 17$$

so that A is the matrix in (6.120). For Jacobi's method, using (6.121) and (6.122)

$$L = D^{-1}(E + F)$$

$$= \begin{bmatrix} 0 & -\dfrac{1}{4} & \dfrac{1}{2} \\ \dfrac{1}{6} & 0 & \dfrac{1}{6} \\ -\dfrac{1}{8} & \dfrac{3}{8} & 0 \end{bmatrix}$$

and from (6.117)

$$\|L\| = \left(\frac{1}{4^2} + \frac{1}{2^2} + \frac{1}{6^2} + \frac{1}{6^2} + \frac{1}{8^2} + \frac{3^2}{8^2} \right)^{1/2}$$
$$= 0.72 < 1$$

so the process converges. From (6.126) with $k = 0$ and $x^{(0)} = 0$ we have

$$x_1^{(1)} = b_1/a_{11} = 4/4 = 1, \qquad x_2^{(1)} = b_2/a_{22} = -9.5/6 = -1.583,$$
$$x_3^{(1)} = b_3/a_{33} = 17/8 = 2.125$$

Next, set $k = 1$ in (6.126) to give

$$x_1^{(2)} = (-a_{12}x_2^{(1)} - a_{13}x_3^{(1)} + b_1)/a_{11}$$
$$= (-x_2^{(1)} + 2x_3^{(1)} + 4)/4 = 2.458$$
$$x_2^{(2)} = (-a_{21}x_1^{(1)} - a_{23}x_3^{(1)} + b_2)/a_{22}$$
$$= (x_1^{(1)} + x_3^{(1)} - 9.5)/6 = -1.062 \qquad (6.129)$$
$$x_3^{(2)} = (-x_1^{(1)} + 3x_2^{(1)} + 17)/8 = 1.406$$

and similarly

$$x^{(3)} = [1.969, -0.939, 1.420]^T, \qquad x^{(4)} = [1.945, -1.018, 1.527]^T,$$
$$x^{(5)} = [2.018, -1.005, 1.500]^T$$

which is converging to the exact solution $x = [2, -1, \frac{3}{2}]^T$. The expressions for $x_i^{(2)}$ have been written out in full to emphasize the simplicity of the method. There is no need to remember any formulae: compute x_1 from the first given equation, x_2 from the second, x_3 from the third, etc., by using the values of the x's found at the previous stage. The matrix representation is necessary only for investigation of convergence.

With the Gauss–Seidel method, using (6.121) and (6.122) it is easy to compute $L = (D - E)^{-1}F$, and (6.117) gives $\|L\| = 0.61 < 1$, so again the process converges. From (6.128) with $k = 0$ and $x^{(0)} = 0$ we obtain

$$x_1^{(1)} = b_1/a_{11} = 4/4 = 1$$
$$x_2^{(1)} = (-a_{21}x_1^{(1)} + b_2)/a_{22} = (x_1^{(1)} - 9.5)/6 = -1.417$$
$$x_3^{(1)} = (-a_{31}x_1^{(1)} - a_{32}x_2^{(1)} + b_3)/a_{33} = (-x_1^{(1)} + 3x_2^{(1)} + 17)/8 = 1.469$$

Next, set $k = 1$ in (6.128) to give

$$x_1^{(2)} = (-a_{12}x_2^{(1)} - a_{13}x_3^{(1)} + b_1)/a_{11}$$
$$= (-x_2^{(1)} + 2x_3^{(1)} + 4)/4 = 2.089$$
$$x_2^{(2)} = (-a_{21}x_1^{(2)} - a_{23}x_3^{(1)} + b_2)/a_{22}$$
$$= (x_1^{(2)} + x_3^{(1)} - 9.5)/6 = -0.990 \qquad\qquad (6.130)$$
$$x_3^{(2)} = (-x_1^{(2)} + 3x_2^{(2)} + 17)/8 = 1.493$$

and similarly

$$x^{(3)} = [1.994, -1.002, 1.500]^T$$

We stress again that the only difference between the Jacobi and Gauss–Seidel methods is that in the latter the latest available values of $x_1, x_2, \ldots, x_{i-1}$ are used to compute x_i – for example, compare the two expressions for $x_2^{(2)}$ in (6.129) and (6.130). In the above example, the Gauss–Seidel method converges to the exact solution more rapidly than Jacobi's method, and this is often, but not always, the case.

Problem 6.42 Find an approximate solution of the equations

$$6x_1 + x_2 - 3x_3 = 17.5$$
$$x_1 + 8x_2 + 3x_3 = 10$$
$$-x_1 + 4x_2 + 12x_3 = -12$$

using the methods of (a) Jacobi, (b) Gauss–Seidel.

Notice that for the matrix A in (6.120) which was used in Example 6.16, in each row the modulus of the diagonal element is greater than the sum of the moduli of the off-diagonal elements, i.e.,

$$|a_{11}| = 4 > |a_{12}| + |a_{13}| = 1 + 2$$
$$|a_{22}| = 6 > |a_{21}| + |a_{23}| = 1 + 1$$
$$|a_{33}| = 8 > |a_{31}| + |a_{32}| = 1 + 3$$

In general, any $n \times n$ matrix A having this property:

$$|a_{ii}| > \sum_{\substack{j=1 \\ j \neq i}}^{n} |a_{ij}|, \qquad i = 1, 2, \ldots, n \qquad\qquad (6.131)$$

is called *diagonal dominant*, and it can be shown that both the Jacobi and Gauss–Seidel methods converge when A in (6.109) has this property (although again this provides only a sufficient condition for convergence). Yet another sufficient condition for convergence of the Gauss–Seidel method is provided by A being real, symmetric, and 'positive definite' (see Section 7.4).

Problem 6.43 If

$$A = \begin{bmatrix} 2 & a & 0 \\ 1 & 3 & b \\ 0 & 1 & 2 \end{bmatrix}$$

where a and b are parameters, compute the matrix L for the Gauss–Seidel procedure and hence prove that the method converges if and only if $|a + b| < 6$. Obtain sufficient conditions for convergence by applying (6.131).

Generally speaking, iterative methods are most useful as compared to direct methods when a very good initial approximation $x^{(0)}$ is known, or when a large proportion of the elements of A is zero (A is then called *sparse*).

6.7.2 *Newton–Raphson type method*

A somewhat different iterative procedure can be developed for calculating A^{-1} by generalizing the Newton–Raphson formula for finding the root of a scalar equation $f(z) = 0$. If z_0 is a sufficiently close initial approximation then (see Barnett and Cronin (1975), Formula 9.8)

$$z_{k+1} = z_k - f(z_k)/f'(z_k), \qquad k = 0, 1, 2, \ldots \qquad (6.132)$$

provides successively better approximations. Let $f(z) = a - z^{-1}$, so the exact root is $z = a^{-1}$. Then $f' = z^{-2}$ and (6.132) becomes

$$z_{k+1} = z_k - (a - z_k^{-1})/z_k^{-2}$$
$$= z_k(2 - az_k) \qquad (6.133)$$

and $z_k \rightarrow a^{-1}$ as $k \rightarrow \infty$. By analogy with (6.133) we define a sequence of matrices $X_1, X_2, X_3, \ldots$ by

$$X_{k+1} = X_k (2I - AX_k), \qquad k = 0, 1, 2, \ldots \qquad (6.134)$$

and show that $X_k \rightarrow A^{-1}$ as $k \rightarrow \infty$. To prove this, define an 'error' matrix

$$E_k = I - AX_k \qquad (6.135)$$

and our task is to show that $E_k \rightarrow 0$ as $k \rightarrow \infty$, for then $AX_k \rightarrow I$ as $k \rightarrow \infty$,

139

which is the desired result. Clearly (6.135) implies

$$E_{k+1} = I - AX_{k+1}$$
$$= I - AX_k(2I - AX_k), \qquad \text{by (6.134)}$$
$$= I - (I - E_k)[2I - (I - E_k)], \qquad \text{by (6.135)}$$
$$= I - (I - E_k)(I + E_k) = E_k^2$$

Hence $E_1 = E_0^2$, $E_2 = E_1^2 = E_0^4$, $E_3 = E_2^2 = E_0^8, \ldots, E_k = E_0^p$, where $p = 2^k$. By the same argument as was applied to L in the preceding section (see (6.115)) it then follows that $E_0^p \to 0$ as $k \to \infty$ provided $E_0 = I - AX_0$ has all its eigenvalues with modulus less than unity. Also, as previously for L in (6.116), a *sufficient* condition for convergence of (6.134) is that $\|E_0\| = \|I - AX_0\| < 1$, and this will be satisfied provided the initial approximation X_0 is sufficiently close to A^{-1}.

Example 6.17 Let

$$A = \begin{bmatrix} 2 & 1 \\ 3 & 4 \end{bmatrix} \tag{6.136}$$

which has the exact inverse

$$A^{-1} = \begin{bmatrix} 0.8 & -0.2 \\ -0.6 & 0.4 \end{bmatrix} \tag{6.137}$$

Take

$$X_0 = \begin{bmatrix} 0.5 & -0.1 \\ -0.3 & 0.2 \end{bmatrix}$$

so that

$$E_0 = I - AX_0 = \begin{bmatrix} 0.3 & 0 \\ -0.3 & 0.5 \end{bmatrix}$$

and from (6.117), $\|E_0\| = 0.66 < 1$, so convergence is assured. From (6.134)

$$X_1 = X_0(2I - AX_0) = \begin{bmatrix} 0.68 & -0.15 \\ -0.45 & 0.30 \end{bmatrix}$$

$$X_2 = X_1(2I - AX_1) = \begin{bmatrix} 0.78 & -0.19 \\ -0.56 & 0.38 \end{bmatrix}$$

which is a reasonable approximation to (6.137).

Problem 6.44 If

$$A = \begin{bmatrix} 5 & 2 \\ 3 & -1 \end{bmatrix}, \qquad X_0 = \begin{bmatrix} 0.1 & 0.2 \\ 0.3 & -0.4 \end{bmatrix}$$

calculate X_1 and X_2 using (6.134), and compare with the exact expression for A^{-1}.

Problem 6.45 If E_k is the matrix defined in (6.135), prove that provided the convergence condition is satisfied then

$$A^{-1} = X_k(I + E_k + E_k^2 + E_k^3 + \cdots)$$

(the sum being taken to infinity).

Problem 6.46 Prove that the sequence defined by

$$Y_{k+1} = (2I - Y_k A)Y_k, \qquad k = 0, 1, 2, \ldots$$

converges to A^{-1} provided all the eigenvalues of $I - Y_0 A$ have modulus less than unity.

Exercises

This chapter is the most important in the book. A fairly large number of exercises is therefore given below, some of which develop further aspects of the theory.

6.1 Let A be a normal matrix (defined in Exercise 2.10) and U be a unitary matrix.
(a) Show that $U^{-1}AU$ is also normal. (It can be shown that every normal matrix is unitarily similar to a *diagonal* matrix.)
(b) If λ is any eigenvalue of U prove that $1/\bar{\lambda}$ is also an eigenvalue of U, and that $|\lambda| = 1$.
(c) Show that $\|Ux\| = \|x\|$ for any vector x.

6.2 If S is an arbitrary $n \times n$ skew hermitian matrix, use the results of Problems 6.11 and 6.15 to show that $I_n + S$ is nonsingular. (This result was used in Exercise 4.11.)

6.3 Use the result of Exercise 4.5 to show that the eigenvalues of the matrix M in (4.78) are the eigenvalues of A together with those of D.

6.4 *Gershgorin's theorem* states that the eigenvalues of an arbitrary $n \times n$ real or complex matrix A lie in the region of the complex z-plane consisting of the n discs having centres a_{ii}, radii $\rho_i = \sum_{\substack{j=1 \\ j \neq i}}^{n} |a_{ij}|$, $i = 1, 2, \ldots, n$ (thus ρ_i is the sum of the moduli of the off-diagonal elements in the ith row of A). The discs can be written $|z - a_{ii}| \leq \rho_i$, $i = 1, \ldots, n$.
   For example, application of the theorem to the matrix A in (6.5) gives the two discs $|z - 1| \leq 3$, $|z - 2| \leq 2$. The eigenvalues lie in the union of these two discs, which in this example is simply the larger disc (the actual values were found in Example 6.1 to be $\lambda_1 = -1$, $\lambda_2 = 4$). The theorem is useful because it enables approximate bounds for the eigenvalues to be determined easily.
   Apply Gershgorin's theorem to the matrices in Problem 6.6, and compare with the actual values of the λ's.

6.5 The definition of a diagonal dominant matrix was given in (6.131). Use Gershgorin's theorem (Exercise 6.4) to prove: (a) any diagonal dominant matrix is nonsingular (use (6.21)); (b) if all the diagonal elements of a diagonal dominant matrix are negative, then all the eigenvalues have negative real parts.

6.6 Let A_n denote an $n \times n$ *tridiagonal* matrix in the form given in Exercise 3.2, so that the matrix in (3.42) is A_4. If ϕ_n is the characteristic polynomial of A_n, expand $\det(\lambda I_n - A_n)$ by the last row to obtain the recurrence formula

$$\phi_n = (\lambda - a_n)\phi_{n-1} - c_{n-1}b_{n-1}\phi_{n-2}, \qquad n = 2, 3, \ldots \qquad (6.138)$$

with $\phi_0 = 1$, $\phi_1 = \lambda - a_1$.

Hence obtain the characteristic polynomial of

$$\begin{bmatrix} 1 & 2 & 0 & 0 \\ 2 & -1 & 1 & 0 \\ 0 & 3 & -3 & -1 \\ 0 & 0 & 7 & 4 \end{bmatrix}$$

6.7 Apply (6.138) to obtain the characteristic equation of the 4×4 matrix in (6.18). By setting $(\lambda - 2)^2 = \mu$, or otherwise, determine the four eigenvalues of this matrix.

6.8 Let ψ_n denote the characteristic polynomial of the $n \times n$ tridiagonal matrix

$$\begin{bmatrix} 0 & 1 & 0 & 0 & & \\ \frac{1}{2} & 0 & \frac{1}{2} & 0 & \mathbf{0} & \\ 0 & \frac{1}{2} & 0 & \frac{1}{2} & & \\ & & & \ddots & & \frac{1}{2} \\ & \mathbf{0} & & & \frac{1}{2} & 0 \end{bmatrix}$$

Use (6.138) to show that if $T_0 = 1$, $T_n = 2^{n-1}\psi_n$ $(n \geq 1)$, then

$$T_n = 2\lambda T_{n-1} - T_{n-2}, \qquad n \geq 2$$

These polynomials $T_0 = 1$, $T_1 = \lambda$, $T_2 = 2\lambda^2 - 1$, $T_3 = 4\lambda^3 - 3\lambda, \ldots$, are called *Chebyshev* polynomials (see Barnett and Cronin (1975), Formula 9.12.1) and are very useful in numerical analysis.

6.9 Using the notation of Exercise 6.6, assume that in (3.42) we have $b_i c_i > 0$, $i = 1, 2, 3$. If

$$D = \text{diag}[1, (b_1/c_1)^{1/2}, (b_1 b_2/c_1 c_2)^{1/2}, (b_1 b_2 b_3/c_1 c_2 c_3)^{1/2}]$$

show that $DA_4 D^{-1}$ is a real symmetric matrix.

This result can be extended to any value of n, showing that when $b_i c_i > 0$, for $i = 1, \ldots, n-1$, then all the eigenvalues of A_n are real, since A_n is similar to a real symmetric matrix.

6.10 Consider the infinitely long resistor ladder network shown in Fig. 6.3. It can be shown that the current i_k in the $(k+1)$th loop satisfies the

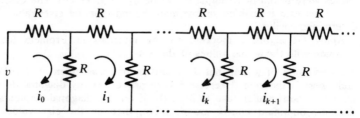

Fig. 6.3 Infinite ladder network for Exercise 6.10.

difference equation

$$i_{k+2} - 3i_{k+1} + i_k = 0, \qquad k = 0, 1, 2, \ldots$$

and $i_k \to 0$ as $k \to \infty$. Define $X_1(k) = i_k$, $X_2(k) = i_{k+1}$, and hence write the equation in the matrix form (6.99). Use (6.101) to show that $i_k = i_0[\frac{1}{2}(3 - \sqrt{5})]^k$.

6.11 If J_n is the matrix defined in Exercise 2.4, prove that the eigenvalues of $J_n A J_n$ are the same as those of A. What are its eigenvectors?

6.12 Application of the preceding exercise to the companion matrix C_n shows that the characteristic polynomial of $J_n C_n J_n$ is $k(\lambda)$ in (6.53), and similarly for C_n^T and $J_n C_n^T J_n$. These three matrices are sometimes used as companion forms. Write out each in full.

6.13 Let the companion matrix C_n defined in Section 6.3.4 be nonsingular. Define an $n \times n$ matrix F_n having the first row

$$[-k_{n-1}/k_n, -k_{n-2}/k_n, \ldots, -k_1/k_n, -1/k_n]$$

all the elements on the diagonal immediately below the principal diagonal equal to unity, and all other elements equal to zero. By calculating the product $C_n F_n$ verify that $F_n = C_n^{-1}$.

Determine the characteristic equation of C_n^{-1} without evaluating $\det(\lambda I - C_n^{-1})$.

6.14 Deduce that the polynomial $k(\lambda)$ in (6.53) has a repeated root if and only if the matrix $D = k'(C_n)$ is singular, where $k'(\lambda) = dk/d\lambda$ ($\det D$ is called the *discriminant* of $k(\lambda)$).

6.15 Prove that if left and right eigenvectors of A are denoted by $v_1, \ldots, v_n$ and $u_1, \ldots, u_n$ respectively, then $v_j u_i = 0$, for $i \neq j$ (assuming as usual that all the eigenvalues of A are distinct). Hence prove that if S is the matrix having rows $v_1, v_2, \ldots, v_n$ and T is the matrix in (6.71), then ST is diagonal.

6.16 Let A be an arbitrary $n \times n$ nonsingular matrix and B any other $n \times n$ matrix. By considering $\det(\lambda I_n - AB) = 0$ and using (4.33), prove that AB and BA have the same eigenvalues. Hence, by applying the first result of Problem 6.15 to the matrix $AB - I_n$ prove that there cannot exist a matrix B such that $AB - BA = I_n$.

6.17 By setting $Z = I_m$ and $W = \lambda I_n$ in Exercise 4.7, prove that

$$\det(\lambda I_n - XY) = \lambda^{n-m} \det(\lambda I_m - YX)$$

where X is an arbitrary $n \times m$ matrix and Y an arbitrary $m \times n$ matrix $(n > m)$.

This shows that the eigenvalues of XY are those of YX together with $n - m$ zeros, thus generalizing the first part of the preceding exercise.

6.18 Construct a real symmetric 3×3 matrix having eigenvalues $\lambda_1 = -2$, $\lambda_2 = 1$, $\lambda_3 = 3$ and corresponding eigenvectors $[1, 2, 2]^T$, $[2, -2, 1]^T$, $[-2, -1, 2]^T$ (hint: use (6.85)).

6.19 Prove that: (a) any idempotent matrix (defined in Exercise 4.14) has its eigenvalues equal to 0 or 1; (b) any nilpotent matrix (defined in Exercise 4.15) has all its eigenvalues equal to zero.

143

6.20 If A is an $n \times n$ matrix having only one nonzero eigenvalue, write down its characteristic equation. Hence show that $\det(I_n + A) = 1 + \text{tr}(A)$.

6.21 If A and B satisfy (6.68) and $p(\lambda)$ is an arbitrary polynomial, prove that $p(B) = P^{-1}p(A)P$. Hence deduce that the minimum polynomials of A and B are identical.

6.22 From (6.96) it is clear that the solution $x(t)$ of the differential equations (6.91) tends to zero as $t \to \infty$ provided each λ_i has negative real part, for then $\exp(\lambda_i t) \to 0$. The system (6.91) is then called *asymptotically stable* (the result still holds even if some λ's are repeated).
If

$$A = \begin{bmatrix} 2 & -1 \\ c & -3 \end{bmatrix}$$

calculate the eigenvalues of A and hence show that in this case the condition for asymptotic stability of (6.91) is $c > 6$.
An alternative method, for application to the characteristic polynomial of A, was given in Exercise 4.2.

6.23 Consider an $n \times n$ matrix having the form

$$A = \begin{bmatrix} a_1 & a_2 & a_3 \ldots a_n \\ a_n & a_1 & a_2 \ldots a_{n-1} \\ a_{n-1} & a_n & a_1 \ldots a_{n-2} \\ \cdot & \cdot & \cdot \quad \cdot \\ \cdot & \cdot & \cdot \quad \cdot \\ \cdot & \cdot & \cdot \quad \cdot \\ a_2 & a_3 & a_4 \ldots a_1 \end{bmatrix}$$

Such matrices are called *circulants*, and arise in a number of applications. By considering the product Au, where $u = [1, \omega, \omega^2, \ldots, \omega^{n-1}]^T$ and ω is an nth root of unity (i.e., $\omega^n = 1$), show that u is an eigenvector of A corresponding to an eigenvalue $a_1 + a_2\omega + a_3\omega^2 + \cdots + a_n\omega^{n-1}$.
The n eigenvalues and eigenvectors of A are obtained by taking the n values of ω.

6.24 Consider two linear control systems in the form (4.36), i.e.,

$$\dot{x}_1 = A_1 x_1 + b_1 u, \qquad \dot{x}_2 = A_2 x_2 + b_2 u$$

where x_1, x_2, b_1, b_2 are column n-vectors. Assume that the corresponding controllability matrices $\mathscr{C}_1$ and $\mathscr{C}_2$ defined in (4.37) are both nonsingular. If the similarity transformation $x_1 = Tx_2$ transforms the first system into the second, prove that

$$Tb_2 = b_1, \qquad TA_2 b_2 = A_1 b_1, \qquad TA_2^2 b_2 = A_1^2 b_1, \ldots$$

and hence show that $T = \mathscr{C}_1 \mathscr{C}_2^{-1}$.

6.25 Using (5.69), apply the method of Section 5.6 to show that the matrix equation

$$AXB - X = C$$

has a unique solution X if and only if $\lambda_i \mu_j \neq 1$, for all i and j (A, B have dimensions $n \times n$ and $m \times m$ and eigenvalues λ_i, μ_j respectively).

6.26 Use the power method with $X_0 = [1, 1, 1]^T$ to estimate the dominant eigenvalue and corresponding eigenvector of

$$A = \begin{bmatrix} 0 & 5 & -6 \\ -4 & 12 & -12 \\ -2 & -2 & 10 \end{bmatrix}$$

6.27 Let P be any nonsingular $n \times n$ matrix such that $Pu_1 = f$, where u_1 is an eigenvector associated with the eigenvalue λ_1 of an $n \times n$ matrix A, and f is the first column of I_n. By considering $PAP^{-1}f$ ($\equiv$ the first column of PAP^{-1}) show that

$$PAP^{-1} = \begin{bmatrix} \lambda_1 & b \\ 0 & A' \end{bmatrix} \begin{matrix} 1 \\ n-1 \end{matrix}$$

(where the elements of b are unimportant). Hence deduce that the eigenvalues of the $(n-1) \times (n-1)$ matrix A' are the remaining $n-1$ eigenvalues of A.

Determine such a P for the matrix A in Exercise 6.26, and by obtaining A' find the other eigenvalues of A.

6.28 If $A = \begin{bmatrix} 1 & 2 \\ 0 & 1 \end{bmatrix}$, show by considering the equation $AP = PD$ with D diagonal, that it is impossible to find a nonsingular P such that A is similar to a diagonal matrix.

Notice that A has eigenvalues 1, 1 but only one independent eigenvector. Further study of such problems involves introduction of the *Jordan form*, which is outside the scope of this book.

6.29 Write a computer program to apply the power method of Section 6.6.1. Test your program on Exercise 6.26.

Given that all the roots of the polynomial

$$\lambda^6 + \lambda^5 - 12\lambda^4 - 4\lambda^3 + 24\lambda^2 + 4\lambda - 8$$

are real and distinct, determine an approximation to the root having largest modulus by applying your program to the companion matrix associated with the polynomial.

6.30 Write a computer program to solve the equations $Ax = b$ using the Gauss–Seidel method, and test your program on Problem 6.42. Use it to solve the equations when

$$A = \begin{bmatrix} B & I_3 & 0 \\ I_3 & B & I_3 \\ 0 & I_3 & B \end{bmatrix}$$

$$b = [-1.08, 0.47, -0.93, -1.16, -0.73, 1.39, -1.68, 0.68, -2.89]^{\mathrm{T}}$$

where

$$B = \begin{bmatrix} -4 & 1 & 0 \\ 1 & -4 & 1 \\ 0 & 1 & -4 \end{bmatrix}$$

Sets of equations having this form arise when using finite difference methods to solve certain partial differential equations.

145

7. Quadratic and hermitian forms

The expression

$$q = ax_1^2 + 2bx_1x_2 + cx_2^2 \tag{7.1}$$

where a, b, c are real constants, is called a *quadratic form* in the two variables x_1 and x_2, since each term has total degree *two* in the x's. We can write (7.1) as

$$q = [x_1, x_2]\begin{bmatrix} ax_1 + bx_2 \\ bx_1 + cx_2 \end{bmatrix}$$

$$= [x_1, x_2]\begin{bmatrix} a & b \\ b & c \end{bmatrix}\begin{bmatrix} x_1 \\ x_2 \end{bmatrix}$$

$$= x^T A x \tag{7.2}$$

The 2×2 matrix A in (7.2) is called the *matrix of the form* q; notice that it is symmetric. Suppose we change coordinates in (7.2) according to $x = Py$. Then $q = (Py)^T A P y = y^T P^T A P y$, and we know from Section 6.4.3 that is is always possible to choose an orthogonal matrix P such that $P^T A P = B$ is *diagonal*. In this case $q = y^T B y$ reduces simply to $b_{11}y_1^2 + b_{22}y_2^2$, a sum of squares. It is interesting to give a simple geometrical interpretation of this result. The ellipse shown in Fig. 7.1a has equation $ax_1^2 + 2bx_1x_2 + cx_2^2 = 1$. Clearly, if we rotated the axes of coordinates anticlockwise through an angle θ, then the ellipse would be in the standard position, shown in Fig. 7.1b, and its equation would then be $y_1^2/\alpha^2 + y_2^2/\beta^2 = 1$ (the reader should confirm that in this case the matrix P is that given in (1.7)).

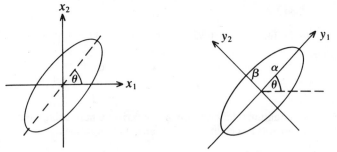

Fig. 7.1a Ellipse $ax_1^2 + 2bx_1x_2 + cx_2^2 = 1$.   **Fig. 7.1b** Ellipse $y_1^2/\alpha^2 + y_2^2/\beta^2 = 1$.

Quadratic forms (and their complex generalization – hermitian forms) arise in many applications (see Section 7.5), and their properties will be developed in this chapter.

7.1 Definitions

The most general quadratic form in n real variables $x_1, \ldots, x_n$ is

$$q = a_{11}x_1^2 + 2a_{12}x_1x_2 + a_{22}x_2^2 + 2a_{13}x_1x_3 + a_{33}x_3^2 + 2a_{23}x_2x_3$$
$$+ \cdots + 2a_{n-1,n}x_{n-1}x_n + a_{nn}x_n^2 \tag{7.3}$$

$$= \sum_{i=1}^{n} \sum_{j=1}^{n} a_{ij}x_ix_j \tag{7.4}$$

In (7.3) the total degree of each term in the x's is two, and the a's are real constants satisfying the condition

$$a_{ij} = a_{ji}, \qquad \text{all } i \neq j \tag{7.5}$$

Generalizing the argument used to obtain (7.2), the form in (7.3) can be written

$$q = [x_1, \ldots, x_n] \begin{bmatrix} a_{11}x_1 + a_{12}x_2 + \cdots + a_{1n}x_n \\ \vdots \\ a_{1n}x_1 + a_{2n}x_2 + \cdots + a_{nn}x_n \end{bmatrix}$$

$$= x^T A x \tag{7.6}$$

where $x = [x_1, x_2, \ldots, x_n]^T$, and $A = [a_{ij}]$ is the $n \times n$ real symmetric matrix of the form. Given q, the matrix A can be written down on noticing that a_{ii} is the coefficient of x_i^2, and a_{ij} $(i \neq j)$ is *one-half* the coefficient of the x_ix_j term. In particular, A being diagonal corresponds to q containing only squared terms.

Example 7.1 For the quadratic form (with $n = 3$)

$$q = 2x_1^2 + 4x_1x_2 + 7x_2^2 + 5x_1x_3 + 6x_2x_3 - x_3^2 \tag{7.7}$$

the matrix of the form is

$$A = \begin{bmatrix} 2 & 2 & \frac{5}{2} \\ 2 & 7 & 3 \\ \frac{5}{2} & 3 & -1 \end{bmatrix} \tag{7.8}$$

Problem 7.1 Write down the $n \times n$ matrix of the form

$$q = 2x_1^2 - 4x_1x_2 + 3x_2^2 - 5x_1x_3 + 10x_3^2$$

(a) when $n = 3$, (b) when $n = 4$.

147

It seems reasonable at this point to ask whether an expression

$$x^T C x \tag{7.9}$$

in which C is an arbitrary (i.e. nonsymmetric) real $n \times n$ matrix, represents a quadratic form more general than (7.6). The answer is 'No', which we demonstrate by splitting up C into its symmetric and skew symmetric parts M and S respectively. From (2.52), $C = M + S$, so that (7.9) becomes

$$x^T C x = x^T M x + x^T S x \tag{7.10}$$

However, since all quadratic forms are *scalar* quantities, in particular transposing $x^T S x$ does not alter its value. Therefore since $S^T = -S$, we obtain

$$x^T S x = (x^T S x)^T = x^T S^T (x^T)^T = -x^T S x$$

which shows that $x^T S x \equiv 0$. Hence in (7.10), $x^T C x \equiv x^T M x$, showing that the form (7.9) is identical to the form associated with the symmetric part $M = \frac{1}{2}(C + C^T)$ of C.

Problem 7.2   Write out in full the form (7.9) when

$$C = \begin{bmatrix} 2 & 1 & 4 \\ 3 & -1 & 2 \\ 1 & -3 & 4 \end{bmatrix}$$

Calculate the symmetric part M of C and verify that $x^T M x$ gives the same expression.

When the variables $x_1, \ldots, x_n$ are allowed to take complex values, the generalization is not to a complex quadratic form (i.e., letting the a's in (7.3) be complex), but instead to the *hermitian form* defined by

$$\begin{aligned} h = a_{11}|x_1|^2 &+ (a_{12}\bar{x}_1 x_2 + \bar{a}_{12} x_1 \bar{x}_2) + a_{22}|x_2|^2 \\ &+ (a_{13}\bar{x}_1 x_3 + \bar{a}_{13} x_1 \bar{x}_3) + \cdots + a_{nn}|x_n|^2 \end{aligned} \tag{7.11}$$

$$= \sum_{i=1}^{n} \sum_{j=1}^{n} a_{ij}\bar{x}_i x_j \tag{7.12}$$

where $|x_1|^2 = \bar{x}_1 x_1$, etc., and the a's are complex numbers satisfying

$$a_{ij} = \bar{a}_{ji}, \qquad \text{for all } i \neq j \tag{7.13}$$

The vector–matrix expression for (7.12) is

$$h = x^* A x \tag{7.14}$$

in which, in view of (7.13), the matrix $A = [a_{ij}]$ is hermitian. The reason for the definition adopted is that h is real, whatever the values of the x's; this is shown by the following sequence of identities:

$$\bar{h} \equiv h^* \equiv (x^*Ax)^* \equiv x^*A^*(x^*)^* \equiv x^*Ax \equiv h$$

For hermitian forms, (7.11) shows that a_{ii} is the coefficient of $|x_i|^2$, and a_{ij} $(i \neq j)$ is the coefficient of $\bar{x}_i x_j$.

Example 7.2 The matrix of the hermitian form

$$h = 3|x_1|^2 + (1+i)\bar{x}_1 x_2 + (1-i)x_1\bar{x}_2 + 2|x_2|^2$$

is

$$A = \begin{bmatrix} 3 & 1+i \\ 1-i & 2 \end{bmatrix}$$

Also,

$$h = 3|x_1|^2 + (1+i)\bar{x}_1 x_2 + \overline{[(1+i)\bar{x}_1 x_2]} + 2|x_2|^2,$$

showing that h is real.

Problem 7.3 Write down the hermitian form corresponding to the hermitian matrix in Example 2.14.

Problem 7.4 If S is a real skew symmetric matrix prove that $x^T S \bar{x}$ is either zero or purely imaginary.

It is worth mentioning here that any hermitian form in n complex variables can be expressed as a quadratic form in $2n$ real variables – see Exercise 7.9.

We now consider the effect of the change of coordinates

$$x = Py \qquad\qquad (7.15)$$

where P is a nonsingular $n \times n$ matrix. For the quadratic form (7.6), substitution of (7.15) gives

$$q = y^T B y \qquad\qquad (7.16)$$

where

$$B = P^T A P \qquad\qquad (7.17)$$

Any matrix B satisfying (7.17) with P nonsingular is said to be *congruent* to A, and the relationship between A and B is called *congruence*. Comparison of (7.17) with (5.14) and (6.68) reveals that congruence is a special case of both equivalence and similarity. In particular, we know from Section 6.4.3 that for *any* real symmetric A, there exists an orthogonal matrix P in (7.17) such that B is the diagonal matrix of the eigenvalues $\lambda_1, \ldots, \lambda_n$ of A. In this case (7.16) becomes simply the sum of squares

$$\lambda_1 y_1^2 + \lambda_2 y_2^2 + \cdots + \lambda_n y_n^2 \tag{7.18}$$

where the λ's are all real.

Example 7.3 We can now reinterpret Example 6.11 in terms of quadratic forms. Corresponding to the matrix A in (6.86) we have

$$q = x_1^2 - 4x_1 x_2 - 2x_2^2 \tag{7.19}$$

which is transformed via

$$\begin{bmatrix} x_1 \\ x_2 \end{bmatrix} = \frac{1}{\sqrt{5}} \begin{bmatrix} 2 & 1 \\ -1 & 2 \end{bmatrix} \begin{bmatrix} y_1 \\ y_2 \end{bmatrix}$$

into $2y_1^2 - 3y_2^2$.

Problem 7.5 Verify the preceding example by directly substituting into (7.19) for x_1 and x_2 in terms of y_1 and y_2.

Problem 7.6 Interpret the result of Problem 6.32 in terms of quadratic forms.

The reader who has attempted Problem 7.5 will appreciate that, even in simple cases, matrices provide a convenient method for manipulating quadratic forms.

For hermitian forms the expression (7.18) is replaced by

$$h = y^*(P^*AP)y$$
$$= \lambda_1 |y_1|^2 + \lambda_2 |y_2|^2 + \cdots + \lambda_n |y_n|^2 \tag{7.20}$$

with P in Eq. (7.15) now a unitary matrix, and $B = P^*AP$ is *conjunctive* to A.

We noted in Section 6.4.3 that the number of nonzero eigenvalues of a real symmetric or hermitian matrix A is equal to $R(A)$ ($= r$, say). Thus in both (7.18) and (7.20) the number of nonzero coefficients is equal to r, which is therefore also called the *rank* of the form.

Problem 7.7 Determine the rank of the quadratic form in Problem 7.1 in both cases (a) and (b).

7.2 Lagrange's reduction of quadratic forms

The preceding method of reducing a quadratic form by orthogonal similarity to a sum of squares requires calculation of eigenvalues and eigenvectors of the matrix of the form. This can be avoided by using

Lagrange's procedure, which is a direct extension of the simple idea of 'completing the square'.

Example 7.4 Consider the following steps which reduce q to a sum of squares:

$$q = 2x_1^2 + 6x_1x_2 + x_2^2 = 2(x_1^2 + 3x_1x_2) + x_2^2$$

$$= 2\left[\left(x_1 + \frac{3}{2}x_2\right)^2 - \frac{9}{4}x_2^2\right] + x_2^2$$

$$= 2\left(x_1 + \frac{3}{2}x_2\right)^2 - \frac{7}{2}x_2^2$$

$$= 2y_1^2 - \frac{7}{2}y_2^2 \qquad (7.21)$$

where $y_1 = x_1 + \frac{3}{2}x_2$, $y_2 = x_2$. In matrix terms

$$\begin{bmatrix} y_1 \\ y_2 \end{bmatrix} = \underbrace{\begin{bmatrix} 1 & \frac{3}{2} \\ 0 & 1 \end{bmatrix}}_{P^{-1}}\begin{bmatrix} x_1 \\ x_2 \end{bmatrix}, \qquad \begin{bmatrix} x_1 \\ x_2 \end{bmatrix} = \underbrace{\begin{bmatrix} 1 & -\frac{3}{2} \\ 0 & 1 \end{bmatrix}}_{P}\begin{bmatrix} y_1 \\ y_2 \end{bmatrix} \qquad (7.22)$$

which gives the transformation matrix P in (7.15).

The procedure in general is as follows:

(a) If $a_{11} \neq 0$, take out a factor a_{11} from all the terms involving x_1, giving $a_{11}(x_1^2 + x_1k)$, where k depends only upon $x_2, x_3, \ldots, x_n$.

(b) Complete the square for these x_1 terms, giving $a_{11}[(x_1 + \frac{1}{2}k)^2 - \frac{1}{4}k^2]$, and let $y_1 = x_1 + \frac{1}{2}k$.

(c) Repeat with the x_2 terms, then the x_3 terms, etc.

As (7.22) shows, the transformation matrix P in (7.15) is not orthogonal in this case.

Example 7.5 We give a further illustration for the case of q in Eq. (7.7).

$$q = 2\left[x_1^2 + x_1\left(2x_2 + \frac{5}{2}x_3\right)\right] + 7x_2^2 + 6x_2x_3 - x_3^2$$

$$= 2\left[\left(x_1 + x_2 + \frac{5}{4}x_3\right)^2 - \frac{1}{4}\left(2x_2 + \frac{5}{2}x_3\right)^2\right] + 7x_2^2 + 6x_2x_3 - x_3^2$$

$$= 2y_1^2 - \frac{1}{2}\left(2x_2 + \frac{5}{2}x_3\right)^2 + 7x_2^2 + 6x_2x_3 - x_3^2$$

$$= 2y_1^2 + 5x_2^2 + x_2x_3 - \frac{33}{8}x_3^2$$

$$= 2y_1^2 + 5\left(x_2^2 + \frac{1}{5}x_2x_3\right) - \frac{33}{8}x_3^2$$

$$= 2y_1^2 + 5\left[\left(x_2 + \frac{1}{10}x_3\right)^2 - \frac{1}{100}x_3^2\right] - \frac{33}{8}x_3^2$$

$$= 2y_1^2 + 5y_2^2 - \frac{167}{40}y_3^2 \qquad (7.23)$$

151

where $y_1 = x_1 + x_2 + \frac{5}{4}x_3$, $y_2 = x_2 + \frac{1}{10}x_3$, $y_3 = x_3$. It is very easy to solve for the x's in terms of the y's, and in matrix form we have

$$\begin{bmatrix} y_1 \\ y_2 \\ y_3 \end{bmatrix} = \underbrace{\begin{bmatrix} 1 & 1 & \frac{5}{4} \\ 0 & 1 & \frac{1}{10} \\ 0 & 0 & 1 \end{bmatrix}}_{P^{-1}} \begin{bmatrix} x_1 \\ x_2 \\ x_3 \end{bmatrix}, \qquad \begin{bmatrix} x_1 \\ x_2 \\ x_3 \end{bmatrix} = \underbrace{\begin{bmatrix} 1 & -1 & -\frac{23}{20} \\ 0 & 1 & -\frac{1}{10} \\ 0 & 0 & 1 \end{bmatrix}}_{P} \begin{bmatrix} y_1 \\ y_2 \\ y_3 \end{bmatrix} \qquad (7.24)$$

Because of the nature of the procedure, it follows that the transformation matrix P^{-1} giving y in terms of x will be upper triangular, with all its diagonal elements unity. Thus (see Section 4.4) P will have the same form, as illustrated in (7.22) and (7.24).

Problem 7.8 Reduce q in Problem 7.1a to a sum of squares by Lagrange's method, stating the transformation obtained.

Example 7.6 If

$$q = x_1^2 + 4x_1x_2 + 4x_2^2 \qquad (7.25)$$

we can immediately see that

$$q = (x_1 + 2x_2)^2 = y_1^2 + 0y_2^2 \qquad (7.26)$$

where $y_1 = x_1 + 2x_2$, $y_2 = x_2$. The transformation is therefore

$$\begin{bmatrix} x_1 \\ x_2 \end{bmatrix} = \begin{bmatrix} 1 & 2 \\ 0 & 1 \end{bmatrix}^{-1} \begin{bmatrix} y_1 \\ y_2 \end{bmatrix} = \begin{bmatrix} 1 & -2 \\ 0 & 1 \end{bmatrix} \begin{bmatrix} y_1 \\ y_2 \end{bmatrix}$$

Notice that it is necessary to define $y_2 = x_2$, so as to ensure the transformation matrix is nonsingular.

The Lagrange procedure detailed above breaks down if $a_{11} = 0$; in this case simply start with x_i, where a_{ii} is the first nonzero diagonal element of A. If all $a_{ii} = 0$ then a preliminary transformation is needed, as in the following example.

Example 7.7 For the form in three variables

$$q = x_1x_2 + 2x_1x_3 - x_2x_3 \qquad (7.27)$$

we first apply the transformation

$$x_1 = z_1, \qquad x_2 = z_1 + z_2, \qquad x_3 = z_3 \qquad (7.28)$$

which introduces a term in z_1^2, i.e.,

152

$$q = z_1(z_1 + z_2) + 2z_1z_3 - (z_1 + z_2)z_3$$
$$= z_1^2 + z_1z_2 + z_1z_3 - z_2z_3 \qquad (7.29)$$

The transformation $x = Pz$ in (7.28) is easily confirmed to be nonsingular, so the rank of q is unaltered. Lagrange's reduction can now be applied to (7.29), giving finally

$$q = y_1^2 - \frac{1}{4}y_2^2 + 2y_3^2 \qquad (7.30)$$

where $y_1 = z_1 + \frac{1}{2}z_2 + \frac{1}{2}z_3 = \frac{1}{2}(x_1 + x_2 + x_3)$, $y_2 = z_2 + 3z_3 = -x_1 + x_2 + 3x_3$, $y_3 = z_3 = x_3$. The choice of transformation in (7.28) is obviously not unique, for example $x_1 = z_1 + z_2$, $x_2 = z_1 - z_2$, $x_3 = z_3$ would produce terms in both z_1^2 and z_2^2.

The above modification of the procedure can be applied at any stage of the reduction when there are no x_i^2 terms.

Interestingly, we shall see in Section 7.4 that Lagrange's reduction can be effectively carried out via gaussian elimination. This avoids the rather tedious manipulations described above.

Problem 7.9 Reduce to a sum of squares each of the following quadratic forms in three variables: (a) $3x_1x_2 + 5x_2x_3 + 2x_3^2$; (b) $x_1x_2 - x_1x_3 + x_2x_3$.

7.3 Sylvester's law of inertia

If a quadratic form $q = x^T Ax$ in n variables is reduced to a sum of squares by *any* nonsingular transformation $x = Py$, then since $B = P^T AP$ in (7.17) is equivalent to A, we have $R(B) = R(A) = r$, so the number of nonzero coefficients in the sum is always equal to r. It is interesting, however, that not only the number of these coefficients remains fixed, but also the distribution of signs. Specifically, suppose q is reduced by two different nonsingular transformations to two sums

$$\alpha_1 y_1^2 + \alpha_2 y_2^2 + \cdots + \alpha_r y_r^2, \qquad \beta_1 z_1^2 + \beta_2 z_2^2 + \cdots + \beta_r z_r^2$$

with all the α's and β's nonzero. Then Sylvester's result is that the number of *positive* α's is equal to the number of positive β's. That is, the numbers π and ν of positive and negative terms in any sum of squares reduction of q remain *constant*, irrespective of the scheme of reduction used – hence the name: law of 'inertia'. The difference $\pi - \nu$ is called the *signature* of q (or of the matrix A).

The same result holds for a hermitian form, except that the reduction is to a sum of squares of moduli, as in Eq. (7.20).

Example 7.8

(a) Consider the form q in (7.19). The transformation $x_1 = 2z_1 + z_2$, $x_2 = z_1$ reduces it to $-6z_1^2 + z_2^2$ so $r = 2$, $\pi = 1$, agreeing with the expression obtained in Example 7.3 using $x_1 = (2y_1 + y_2)/\sqrt{5}$, $x_2 = (-y_1 + 2y_2)/\sqrt{5}$.
(b) For the form in (7.25), the transformation $x_1 = -2z_1 + z_2$, $x_2 = z_1 + 2z_2$ reduces it to $0z_1^2 + 25z_2^2$, so $r = 1$, $\pi = 1$, agreeing with (7.26).
(c) For the form in (7.27), the transformations $x_1 = z_1 + z_2$, $x_2 = z_1 - z_2$, $x_2 = z_3$, and then $z_1 = w_1 - \frac{1}{2}w_3$, $z_2 = w_2 + \frac{3}{2}w_3$, $z_3 = w_3$ produce $w_1^2 - w_2^2 + 2w_3^2$, so $r = 3$, $\pi = 2$, agreeing with (7.30).

Problem 7.10 What are the rank and signature of the quadratic form in Problem 7.9b? Reduce this form by starting with a different choice of z's from the one you used previously, and hence confirm Sylvester's law in this case.

7.4 Sign properties

Of particular importance in many applications are forms which do not change sign, whatever the values of the variables.

7.4.1 Definitions

The following definitions apply equally to both hermitian and quadratic forms.

A form $p(x)$ is called *positive definite* if it is positive everywhere except at the origin, i.e., $p(x) > 0$ for all $x \neq 0$, and $p(0) = 0$; $p(x)$ is *positive semidefinite* if $p(x) \geq 0$ for all x, with at least one $x \neq 0$ such that $p(x) = 0$. Similarly, p is *negative* definite or semidefinite if $p(x) < 0$, or $p(x) \leq 0$, respectively. If $p(x)$ can take both negative and positive values then it is called *indefinite*. These various terms describe the *sign property* of a form.

Example 7.9

(a) If $n = 3$ the form $q = 2x_1^2 + x_2^2 + 3x_3^2$ is clearly positive definite; however, if $n = 4$ then we have $q = 2x_1^2 + x_2^2 + 3x_3^2 + 0x_4^2$, so $q = 0$ when $x_1 = x_2 = x_3 = 0$ but $x_4 \neq 0$, and hence q is positive semidefinite.
(b) The form $q = -x_1^2 - 2x_1x_2 - x_2^2 = -(x_1 + x_2)^2$ is negative semidefinite, since $q \leq 0$, and $q = 0$ when $x_1 = -x_2 \neq 0$.
(c) The form $q = 2x_1^2 + x_2^2 - 3x_3^2$ is indefinite, since for example, when $x_1 = 1$, $x_2 = x_3 = 0$ then $q > 0$, and when $x_3 = 1$, $x_1 = x_2 = 0$ then $q < 0$.
(d) The form $q = x_1x_2 + x_1x_3 + x_2x_3$ is indefinite, since, for example, taking $x_1 = \pm 1$, $x_2 = 1$, $x_3 = 0$ shows that q can take both positive and negative values.

The various terms are also applied to the matrix of the form. It is worth recalling two remarks made in previous chapters concerning a real symmetric matrix A being positive definite: in Section 3.3 we noted that it ensures that there exists a real nonsingular triangular matrix U such that $A = U^T U$; and in Section 6.7.1 we stated that it provides a sufficient condition for convergence of the Gauss–Seidel method for solving $Ax = b$.

Problem 7.11 What is the sign nature of the form $q = (x_1 + x_2 + x_3)^2$ (with $n = 3$)?

Problem 7.12 If A and B are two positive definite hermitian $n \times n$ matrices, prove that $A + B$ is also positive definite.

Problem 7.13 Generalize the result of Example 7.9d by showing that q in Eq. (7.4) is indefinite if all the coefficients a_{ii}, $i = 1, 2, \ldots, n$, are zero.

7.4.2 Tests

We now consider how to determine the sign property of a form. The first step is to notice that the transformation $x = Py$ applied to any form does not alter its sign property, provided P is nonsingular. This is because, as the components of x vary through all possible values so do those of y, with $y = 0$ if and only if $x = 0$. In particular, suppose P is chosen so that a quadratic form $q = x^T A x$ is reduced to the sum of squares

$$\lambda_1 y_1^2 + \lambda_2 y_2^2 + \cdots + \lambda_r y_r^2 \tag{7.31}$$

where as before $r = R(A)$ and the λ's are the eigenvalues of A. Then the sign properties of q are the same as those of (7.31). Since the λ's in (7.31) are all real, then it is obvious that (7.31) will be positive for all $y \neq 0$ only if $r = n$ and each λ is positive – otherwise it would be possible to make (7.31) zero or negative (compare with Example 7.9). Other cases can be argued similarly, to obtain the following:

A quadratic or hermitian form is:
positive definite if $\lambda_i > 0$, $i = 1, 2, \ldots, n$;
negative definite if $\lambda_i < 0$, $i = 1, 2, \ldots, n$;
positive (negative) semidefinite if $\lambda_i \geq 0$ $(\lambda_i \leq 0)$, $i = 1, 2, \ldots, n$, with at least one λ equal to zero;
indefinite if there is at least one positive and one negative λ.

Because of the law of inertia, the preceding conditions on the λ's also apply to the coefficients in *any* sum-of-squares reduction obtained by a

155

nonsingular transformation. We concentrate on quadratic forms, since these are most common in applications.

Example 7.10 We can now determine the sign properties of the forms which we earlier reduced to sums of squares by Lagrange's method. In Example 7.4 there is one positive and one negative coefficient in the reduced form (7.21), so q is indefinite – for example, when $y_1 = 1$, $y_2 = 0$ (corresponding from (7.22) to $x_1 = 1$, $x_2 = 0$) then $q > 0$; and when $y_1 = 0$, $y_2 = 1$ (corresponding to $x_1 = -\frac{3}{2}$, $x_2 = 1$) then $q < 0$.

Similarly, the form in Example 7.5 is indefinite, by inspection of (7.23). However, the form in Example 7.6 is positive semidefinite, since in (7.26) there is one positive and one zero coefficient.

Problem 7.14 Determine the sign properties of the quadratic forms in Problems 7.8 and 7.9.

Problem 7.15 Prove that the trace and determinant of a positive definite hermitian (or real symmetric) matrix are both positive.

Problem 7.16 Show that the condition for a real symmetric 2×2 matrix A to be positive definite is $a_{11} > 0$, $\det A > 0$.

Problem 7.17 Deduce from the result in Problem 7.12 that if two $n \times n$ hermitian matrices A and B have all their eigenvalues positive, then so does $A + B$.

Problem 7.18 If A is a positive definite real symmetric matrix, prove that A^p has the same sign property for any positive integer p. Does this hold for any other values of p?

Example 7.10 shows that the sign property of a quadratic form can be determined immediately from the Lagrange sum of squares. It is now appropriate to show how Lagrange's method can be carried out via gaussian elimination. This is best done by a numerical example: consider the form q in Eq. (7.7) which has matrix A in (7.8). Applying gaussian elimination without partial pivoting (i.e., without row interchanges) to A gives

$$A \to \begin{bmatrix} 2 & 2 & \frac{5}{2} \\ 0 & 5 & \frac{1}{2} \\ 0 & \frac{1}{2} & -\frac{33}{8} \end{bmatrix} \to \begin{bmatrix} 2 & 2 & \frac{5}{2} \\ 0 & 5 & \frac{1}{2} \\ 0 & 0 & -\frac{167}{40} \end{bmatrix} \tag{7.32}$$

Two interesting points are apparent in the final reduced form in (7.32). First, the pivots are identical to the coefficients in the Lagrange sum-of-squares reduction (7.23) of q. Second, if each row is divided by the pivot in that row, then the resulting matrix is precisely P^{-1} in (7.24), giving the transformation $y = P^{-1}x$. It can be shown that these results hold in general: Apply gaussian elimination without partial pivoting to the matrix A of the form q; provided there are no zero pivots, then the coefficients in the Lagrange reduction of q to a sum of squares are given by the pivots in the triangularized array, and if each row in this array is divided by its pivot then the resulting matrix is P^{-1} in the Lagrange transformation $y = P^{-1}x$.

It follows that for q to be positive (negative) definite all the pivots must be positive (negative). If pivots having opposite signs are encountered then q is indefinite and the elimination process can be stopped. If a zero pivot occurs then q is not definite.

It is worth noting that after each step of the elimination procedure the remaining non-triangular block in the bottom right-hand corner is still symmetric, as illustrated by the first array in (7.32) (see Exercise 7.6).

Example 7.11 If $q = x_1^2 + 4x_1x_2 + 6x_2^2 + 2x_1x_3 + 2x_2x_3 + 5x_3^2$ then

$$A = \begin{bmatrix} 1 & 2 & 1 \\ 2 & 6 & 1 \\ 1 & 1 & 5 \end{bmatrix} \rightarrow \begin{bmatrix} 1 & 2 & 1 \\ 0 & 2 & -1 \\ 0 & 0 & \frac{7}{2} \end{bmatrix}$$

after elimination. The pivots are all positive, showing that q and A are positive definite. From the triangularized array, Lagrange's reduction of q would give $y_1^2 + 2y_2^2 + \frac{7}{2}y_3^2$, where $y_1 = x_1 + 2x_2 + x_3$, $y_2 = x_2 - \frac{1}{2}x_3$, $y_3 = x_3$ (the reader should confirm this).

Problem 7.19 Test the form in Problem 7.1a for definiteness using gaussian elimination. Compare your result with the Lagrange reduction obtained in Problem 7.8.

If a zero pivot occurs, but it is required to determine the precise sign property of the form, then there are three different possibilities. The simplest is if $R(A) = r < n$, and all the first r pivots are nonzero. Then the correspondence with Lagrange's reduction still holds, so q is positive (negative) semidefinite if all the nonzero pivots are positive (negative). If, however, a zero pivot occurs at some stage (the ith, say, with $i \leq r$) then a little more care is needed: First, suppose that there is a *subsequent* nonzero diagonal element in some row k. The operations $(Ri) \leftrightarrow (Rk)$ *and* $(Ci) \leftrightarrow (Ck)$ bring this nonzero element into the ith diagonal position, and the triangularization can then be continued.

Notice that it is necessary to carry out both row and column interchanges, so as to preserve the symmetry. This is equivalent to interchanging x_i and x_k in q, so does not affect the sign property. Finally, if no such k exists, then all the remaining diagonal elements are zero, and q in this case is indefinite, as the following argument shows. The reduced form of A at the ith stage is

$$\begin{matrix} i-1 \\ \\ i-1 \end{matrix} \left[\begin{array}{c|c} B & C \\ \hline 0 & E \end{array} \right]$$

where $B = [b_{st}]$ is upper triangular with all diagonal elements b_{ss} nonzero and $E = [e_{pq}]$ has all diagonal elements e_{pp} equal to zero. This corresponds to

$$q = b_{11}y_1^2 + b_{22}y_2^2 + \cdots + b_{i-1,i-1}y_{i-1}^2 + \sum_{\substack{p=i \\ p \neq q}}^{n} \sum_{q=i}^{n} e_{pq}x_p x_q \qquad (7.33)$$

where $y_s = (b_{ss}x_s + b_{s,s+1}x_{s+1} + \cdots + c_{sn}x_n)/b_{ss}$. Setting $y_1 = 0$, $y_2 = 0, \ldots,$ $y_{i-1} = 0$ gives $i - 1$ homogeneous equations in the x's, which can always be solved for $x_1, \ldots, x_{i-1}$ in terms of $x_i, \ldots, x_n$. The form q then reduces to the last term in (7.33), which since it contains no squares shows that q is indefinite (see Problem 7.13).

Example 7.12 We give an example of each of the three cases just discussed.
(a) For q in (7.25),

$$A = \begin{bmatrix} 1 & 2 \\ 2 & 4 \end{bmatrix} \rightarrow \begin{bmatrix} 1 & 2 \\ 0 & 0 \end{bmatrix}$$

showing that q and A are positive semidefinite (agreeing with the sum of squares (7.26)).
(b) If $q = x_1^2 + 4x_1x_2 + 4x_2^2 + 6x_1x_3 + 14x_2x_3 + 10x_3^2$, then

$$A = \begin{bmatrix} 1 & 2 & 3 \\ 2 & 4 & 7 \\ 3 & 7 & 10 \end{bmatrix} \rightarrow \begin{bmatrix} 1 & 2 & 3 \\ 0 & 0 & 1 \\ 0 & 1 & 1 \end{bmatrix} \xrightarrow[\text{(C2)} \leftrightarrow \text{(C3)}]{\text{(R2)} \leftrightarrow \text{(R3)}}$$

$$\begin{bmatrix} 1 & 3 & 2 \\ 0 & 1 & 1 \\ 0 & 1 & 0 \end{bmatrix} \rightarrow \begin{bmatrix} 1 & 3 & 2 \\ 0 & 1 & 1 \\ 0 & 0 & -1 \end{bmatrix}$$

showing that q and A are indefinite.
(c) If $q = x_1^2 + 4x_1x_2 + 4x_2^2 - 2x_1x_3 + x_3^2$, then

$$A = \begin{bmatrix} 1 & 2 & -1 \\ 2 & 4 & 0 \\ -1 & 0 & 1 \end{bmatrix} \rightarrow \begin{bmatrix} 1 & 2 & -1 \\ 0 & 0 & 2 \\ 0 & 2 & 0 \end{bmatrix}$$

No further nonzero pivots can be obtained, showing that q and A are indefinite (the reduced form of A corresponds to $q = (x_1 + 2x_2 - x_3)^2 + 4x_2x_3$, which becomes $4x_2x_3$ on taking $x_1 = -2x_2 + x_3$).

Problem 7.20 Use gaussian elimination to test the following forms for definiteness: (a) $-x_1^2 + 2x_1x_2 - 2x_2^2 + 4x_2x_3 - 4x_3^2$; (b) the form of Problem 7.9a.

Problem 7.21 Determine the sign properties of the following forms using gaussian elimination:

(a) $x_1^2 + 4x_1x_2 + 4x_2^2 + 6x_1x_3 - 4x_2x_3 + 9x_3^2$
(b) $x_1^2 + 2x_1x_2 + x_2^2 + 4x_1x_3 + 4x_2x_3 + 5x_3^2$
(c) $x_1^2 + 11x_2^2 + 13x_3^2 + 14x_4^2 + 2x_1x_2 - 6x_1x_3 + 4x_1x_4 + 10x_2x_4 - 10x_2x_3 - 2x_3x_4$

Problem 7.22 Prove that if A and B are each positive definite hermitian matrices, then so is $A \otimes B$ (hint: consider eigenvalues). If A and B have the same dimensions, does the result hold for the ordinary product AB?

To close this section, we note that the leading principal minors of a positive definite matrix are all positive, since they are equal to successive products of pivots, which are themselves all positive. This is often stated as a theoretical test for definiteness.

Problem 7.23 What are the sign conditions on leading principal minors of a negative definite matrix?

7.5 Some applications

7.5.1 Geometrical interpretation

As was seen at the beginning of this chapter, if A is a real symmetric 2×2 matrix then

$$x^T A x = 1 \qquad (7.34)$$

is the equation of a curve in two dimensions. Applying an orthogonal transformation $x = Py$ to (7.34), which is equivalent to rotating the curve in the plane, reduces it to $\lambda_1 y_1^2 + \lambda_2 y_2^2 = 1$, where λ_1 and λ_2 are the eigenvalues of A. It is clear from this reduced form that if λ_1 and λ_2 are both positive (i.e., A is positive definite) then (7.34) represents an ellipse; if λ_1 and λ_2 have opposite signs (i.e., A is indefinite) then (7.34) represents a hyperbola; and if one eigenvalue is positive and the other zero, then A is positive semidefinite and (7.34) represents a pair of straight lines.

When A is 3×3 then (7.34) describes a surface in three dimensions. For example, if A is positive definite then the reduced form $\lambda_1 y_1^2 + \lambda_2 y_2^2 + \lambda_3 y_3^2 = 1$ shows that (7.34) represents an ellipsoid; in particular, if $\lambda_1 = \lambda_2 = \lambda_3 > 0$ this becomes a sphere. If $\lambda_1 = \lambda_2 > 0$, $\lambda_3 < 0$, the surface is called a 'rotational hyperboloid', and is the familiar shape of an electricity generating station's cooling tower.

Problem 7.24 Prove that the plane curve $ax_1^2 + bx_1x_2 + cx_2^2 = 1$, with $a > 0$, is an ellipse if $b^2 < 4ac$ and a hyperbola if $b^2 > 4ac$.

7.5.2 Optimization of functions

Let $\phi(x) \equiv \phi(x_1, x_2, \ldots, x_n)$ be a real scalar function of n real variables, having continuous first and second partial derivatives with respect to all the variables. The vector of partial derivatives $[\partial\phi/\partial x_1, \partial\phi/\partial x_2, \ldots, \partial\phi/\partial x_n]^T$ is called the *gradient* of ϕ, and is written $\nabla\phi$ or grad ϕ. It is a standard result that for there to be a local maximum or minimum at some point $x = a$ then all the first partial derivatives must be zero there, i.e.,

$$\nabla\phi = 0, \qquad \text{at } x = a \qquad (7.35)$$

Next, define the constant elements h_{ij} of the $n \times n$ *hessian matrix* H (named after Hesse):

$$h_{ij} = \left(\frac{\partial^2\phi}{\partial x_i \partial x_j}\right)_{x=a}, \qquad i, j = 1, \ldots, n \qquad (7.36)$$

(notice that $h_{ij} = h_{ji}$, since by assumption of continuity $\partial^2\phi/\partial x_i\partial x_j = \partial^2\phi/\partial x_j\partial x_i$). Then $x = a$ is a local minimum if H is positive definite, and a local maximum if H is negative definite. The method of proving this result is to expand $\phi(a + \delta x)$, where $\delta x = [\delta x_1, \ldots, \delta x_n]^T$ using Taylor's theorem. In view of (7.35) the terms which are linear in δx vanish, leaving

$$\phi(a + \delta x) - \phi(a) = \frac{1}{2}(\delta x)^T H(\delta x) + \left(\begin{array}{l}\text{cubic and higher-}\\ \text{degree terms in } \delta x\end{array}\right) \qquad (7.37)$$

where H is the matrix defined in (7.36). For $x = a$ to be a local minimum we require $\phi(a) < \phi(a + \delta x)$ for *all* small variations $\delta x \neq 0$. Thus the right-hand side of (7.37) must be positive, and this will be the case for sufficiently small δx if H is positive definite.

Example 7.13 When $n = 2$ the condition for H to be positive definite is $h_{11} > 0$, $\det H > 0$ (see Problem 7.16) which produces

$$\frac{\partial^2\phi}{\partial x_1^2} > 0, \qquad \left(\frac{\partial^2\phi}{\partial x_1^2}\right)\left(\frac{\partial^2\phi}{\partial x_2^2}\right) - \left(\frac{\partial^2\phi}{\partial x_1\partial x_2}\right)^2 > 0$$

(all derivatives being evaluated at $x = a$) and this is the well-known condition for a point $x = a$, at which $\partial\phi/\partial x_1 = \partial\phi/\partial x_2 = 0$, to be a local minimum for $\phi(x_1, x_2)$.

It is useful to note the case when ϕ is itself a quadratic form, i.e. $\phi = x^T A x$. From the full expression in (7.3) it is obvious that

$$\partial\phi/\partial x_1 = 2a_{11}x_1 + 2a_{12}x_2 + 2a_{13}x_3 + \cdots + 2a_{1n}x_n$$
$$\vdots$$
$$\partial\phi/\partial x_n = 2a_{n1}x_1 + 2a_{n2}x_2 + \cdots + 2a_{nn}x_n$$

so in vector form we have

$$\nabla\phi = 2Ax \tag{7.38}$$

Problem 7.25 If $\phi(x) = \frac{1}{2}x^T Q x - x^T b$, where Q is a positive definite real symmetric $n \times n$ matrix and b is a constant column n-vector, show that ϕ has a local minimum at $x = Q^{-1}b$.

Problem 7.26 If u and v are two scalar differentiable functions of $x_1, \ldots, x_n$, prove that $\nabla(uv) = v(\nabla u) + u(\nabla v)$.

7.5.3 Rayleigh quotient

If A is a nonsingular hermitian matrix, then from (7.20)

$$h = x^* A x = \lambda_1 |y_1|^2 + \cdots + \lambda_n |y_n|^2$$

where $x = Py$ with P unitary. Let the eigenvalues of A be numbered so that $\lambda_1 \geq \lambda_2 \geq \cdots \geq \lambda_n$. It then follows that

$$\lambda_n |y_1|^2 + \lambda_n |y_2|^2 + \cdots + \lambda_n |y_n|^2 \leq h \leq \lambda_1 |y_1|^2 + \lambda_1 |y_2|^2 + \cdots + \lambda_1 |y_n|^2$$

i.e.

$$\lambda_n(y^*y) \leq h \leq \lambda_1(y^*y) \tag{7.39}$$

Because P is unitary, $x^*x = y^*P^*Py = y^*y$, so (7.39) implies that $\lambda_n(x^*x) \leq h \leq \lambda_1(x^*x)$, for arbitrary $y \neq 0$. Finally, since $x^*x \neq 0$ if $x \neq 0$, we can divide in (7.39) to obtain

$$\lambda_n \leq \frac{x^* A x}{x^* x} \leq \lambda_1 \tag{7.40}$$

for arbitrary $x \neq 0$. The ratio $r = x^* A x / x^* x$ ($= x^T A x / x^T x$ when A is real symmetric) is called the *Rayleigh quotient*. From (7.40) we have $\lambda_1 \geq r$, $\lambda_n \leq r$, which provides an easy way of estimating bounds for the largest and smallest eigenvalues of A, as the following example demonstrates.

Example 7.14 For the quadratic form in (7.19) the Rayleigh quotient is

$$r = \frac{x_1^2 - 4x_1x_2 - 2x_2^2}{x_1^2 + x_2^2} \tag{7.41}$$

Some choices for x_1 and x_2, and the corresponding values of r, are:

$$x^T = [1, 0], \quad r = 1; \qquad x^T = [0, 1], \quad r = -2;$$
$$x^T = [1, 1], \quad r = -2.5; \qquad x^T = [1, -1], \quad r = 1.5.$$

Since the largest and smallest of these values of r are 1.5 and -2.5 respectively, we have $\lambda_1 \geqslant 1.5$, $\lambda_2 \leqslant -2.5$. The exact values of λ_1 and λ_2 are 2 and -3 respectively (see Example 6.11).

Problem 7.27 Obtain bounds for the largest and smallest eigenvalues of the matrix in (6.87) using the Rayleigh quotient.

7.5.4 Liapunov stability

We now return to the system of linear differential equations discussed in Section 6.5, i.e.,

$$\dot{x} = Ax \tag{7.42}$$

where A is a real constant $n \times n$ matrix. As stated in Exercise 6.22, the system is called asymptotically stable if the solution $x(t)$ of (7.42) tends to zero as $t \to \infty$, a necessary and sufficient condition for this being that all the eigenvalues of A have negative real parts. A method due to the Russian mathematician Liapunov avoids calculation of the eigenvalues. It can be shown that (7.42) is asymptotically stable if a positive definite quadratic form $V = x^T Px$ can be found such that its derivative with respect to time is negative definite. This derivative is, by the product rule,

$$dV/dt = \dot{x}^T Px + x^T P\dot{x}$$

since P is a constant matrix, and substituting for $\dot{x}$ from (7.42) we have

$$dV/dt = (Ax)^T Px + x^T P(Ax)$$
$$= x^T A^T Px + x^T PAx$$
$$= x^T(A^T P + PA)x = x^T Qx$$

showing that dV/dt is also a quadratic form, whose matrix

$$Q = A^T P + PA \tag{7.43}$$

is required to be negative definite. In practice the method is to make a simple choice for Q, e.g., $-I_n$, and then solve (7.43) for P and test the sign property of this solution.

Problem 7.28 Verify that Q in (7.43) is symmetric.

Problem 7.29 In (7.43) take $Q = -I_2$,

$$A = \begin{bmatrix} -1 & -1 \\ 2 & -4 \end{bmatrix}, \qquad P = \begin{bmatrix} p_1 & p_2 \\ p_2 & p_3 \end{bmatrix}$$

and solve for the elements of P. Hence determine whether (7.42) is asymptotically stable in this case.

Problem 7.30 The equation (7.43) is a special case of (5.51). Use the criterion developed in Section 6.3.5 to determine the condition which A must satisfy in order that the solution P of (7.43) be unique.

Exercises

7.1 If A is real, symmetric, and nonsingular, prove that $x^T A x$ and $x^T A^{-1} x$ have the same signature.

7.2 By considering the quadratic form $x^T A^T A x$, prove that if A is a real nonsingular $n \times n$ matrix then $A^T A$ is positive definite. Prove also the converse, that any positive definite real symmetric matrix can be expressed as $B^T B$ with B nonsingular.
   What is the sign property of $A^T A$ if A is $m \times n$?

7.3 If $B = A^T A - A A^T$, where A is any real $n \times n$ matrix such that $B \neq 0$, show that B is symmetric. Use Exercise 2.7 and Eq. (6.23) to prove that B is indefinite.

7.4 Determine the range of values of k for which each of the following forms is positive definite:

(a) $x_1^2 + 4x_1 x_2 + (7 - k)x_2^2 + 8x_1 x_3 + (46 - 10k)x_2 x_3 + (89 - 24k)x_3^2$
(b) $x_1^2 + 6x_1 x_2 + 8x_1 x_3 + k(x_2^2 + x_3^2)$

7.5 Use Gershgorin's theorem (given in Exercise 6.4) to prove that if a real symmetric matrix $A = [a_{ij}]$ has $a_{ii} > 0$ for all i, and is diagonal dominant (defined in (6.131)), then A is positive definite.

7.6 If A is a real symmetric $n \times n$ matrix with $a_{11} \neq 0$, and the first column is reduced in the usual way by gaussian elimination, prove that the last $n - 1$ rows and columns of the resulting array also form a symmetric matrix.

7.7 If A is real and symmetric and $\det A < 0$, prove there exists a real vector x such that $x^T A x < 0$ (hint: use (6.1) and (6.21)).
   Find such an x for the case when A is the matrix in (6.87).

7.8 If A is an arbitrary real symmetric matrix, prove that there exists a real number c such that $A + cI$ is positive definite.

7.9 In the hermitian form $h = x^* A x$, let $A = A_1 + i A_2$, $x = u + iv$, with A_1, A_2, u, v purely real. Prove that $h = u^T A_1 u + v^T A_1 v - 2u^T A_2 v$, and hence show that $h = y^T D y$, where D is the real symmetric $2n \times 2n$ matrix in (2.76) and $y^T = [u^T, -v^T]$.

163

7.10 Let A be an arbitrary positive definite real symmetric $n \times n$ matrix. Apply the definition of $A^{1/2}$ given in (6.77), with T orthogonal, to prove that $A^{1/2}$ is also real symmetric.

If B is any real symmetric $n \times n$ matrix, by considering the product $(A^{1/2})^{-1}ABA^{1/2}$, prove that all the eigenvalues of AB are real.

7.11 For the Rayleigh quotient r associated with a real symmetric matrix A, use the product rule in Problem 7.26 and Eq. (7.38) to show that

$$\nabla r = 2Ax/(x^{\mathrm{T}}x) - 2(x^{\mathrm{T}}Ax)x/(x^{\mathrm{T}}x)^2$$

Hence show that critical points of r (where $\nabla r = 0$) are eigenvectors of A, and associated critical values of r are eigenvalues of A. Verify this for r in (7.41) by direct differentiation (see Example 6.11 for eigenvalues and eigenvectors in this case).

7.12 Write a computer program to determine the sign property of a quadratic form in at most four variables, using gaussian elimination. Test your program on the forms of Problem 7.21. Use it to determine the sign property of the form having matrix

$$\begin{bmatrix} 0.84 & 0.40 & -0.42 & 0.36 \\ 0.40 & 0.62 & 1.16 & 0.01 \\ -0.42 & 1.16 & 4.79 & 0.48 \\ 0.36 & 0.01 & 0.48 & 5.38 \end{bmatrix}$$

8. Introduction to matrix functions

So far in this book we have mainly been concerned with algebraic manipulations of matrices, and have encountered limiting processes only in relation to iterative methods discussed in Chapter 6. We now give a brief introduction to what can be called 'matrix analysis', where the idea of convergence plays a fundamental role.

8.1 Definition and properties

Let

$$f(\lambda) = f_0 + f_1\lambda + f_2\lambda^2 + f_3\lambda^3 + \cdots \tag{8.1}$$

be a scalar infinite power series convergent for $|\lambda| < R$. Consider the matrix series obtained by replacing λ by an $n \times n$ matrix A:

$$f(A) = f_0 I_n + f_1 A + f_2 A^2 + f_3 A^3 + \cdots \tag{8.2}$$

Assuming that all the eigenvalues $\lambda_1, \ldots, \lambda_n$ of A are distinct, we can use the diagonalization formula (6.74), i.e., $A = T\Lambda T^{-1}$, which together with (6.76), namely $A^k = T\Lambda^k T^{-1}$, reduces (8.2) to

$$f(A) = f_0 I + f_1 T\Lambda T^{-1} + f_2 T\Lambda^2 T^{-1} + f_3 T\Lambda^3 T^{-1} + \cdots$$
$$= T(f_0 I + f_1\Lambda + f_2\Lambda^2 + \cdots)T^{-1} \tag{8.3}$$

Since $\Lambda = \text{diag}[\lambda_1, \lambda_2, \ldots, \lambda_n]$ it follows that the matrix within brackets in (8.3) is also diagonal, and the i,i element is $f_0 + f_1\lambda_i + f_2\lambda_i^2 + \cdots$, which will converge to $f(\lambda_i)$ if $|\lambda_i| < R$, and diverge if $|\lambda_i| > R$. Thus, provided *all* the eigenvalues of A lie within the radius of convergence of the series (8.1), we can write (8.3) as

$$f(A) = T \, \text{diag}[f(\lambda_1), f(\lambda_2), \ldots, f(\lambda_n)]T^{-1} \tag{8.4}$$

In this case it therefore makes sense to use (8.2) as the definition of the *matrix function $f(A)$*. (It can be shown that this definition still holds even if some of the eigenvalues of A are repeated.)

Example 8.1 It is well known that

$$\ln(1 + \lambda) = \lambda - \frac{1}{2}\lambda^2 + \frac{1}{3}\lambda^3 - \cdots$$

165

provided $|\lambda| < 1$. Thus we can write

$$\ln(I_n + A) = A - \frac{1}{2}A^2 + \frac{1}{3}A^3 - \cdots \qquad (8.5)$$

provided all the eigenvalues of A have modulus less than unity. For example, if

$$A = \begin{bmatrix} 0 & 1 \\ -\frac{1}{8} & \frac{3}{4} \end{bmatrix} \qquad (8.6)$$

then $\lambda_1 = \frac{1}{4}$, $\lambda_2 = \frac{1}{2}$ and the convergence criterion is satisfied, so (8.5) holds. In fact (8.6) is in companion form, so from (6.88) the matrix transforming A to diagonal form is given by

$$T = \begin{bmatrix} 1 & 1 \\ \lambda_1 & \lambda_2 \end{bmatrix} = \begin{bmatrix} 1 & 1 \\ \frac{1}{4} & \frac{1}{2} \end{bmatrix}, \qquad T^{-1} = \begin{bmatrix} 2 & -4 \\ -1 & 4 \end{bmatrix} \qquad (8.7)$$

Hence from (8.4) we have

$$\ln(I_2 + A) = T \operatorname{diag}\left[\ln\left(1 + \frac{1}{4}\right), \ln\left(1 + \frac{1}{2}\right)\right]T^{-1}$$

which becomes, on using (8.6) and (8.7)

$$\ln\begin{bmatrix} 1 & 1 \\ -\frac{1}{8} & \frac{7}{4} \end{bmatrix} = \begin{bmatrix} \ln\frac{25}{24} & 4\ln\frac{6}{5} \\ \frac{1}{2}\ln\frac{5}{6} & \ln\frac{9}{5} \end{bmatrix} \qquad (8.8)$$

If (8.1) converges for all finite values of λ, then (8.2) converges for *all* $n \times n$ matrices having finite elements. For example,

$$\sin A = A - \frac{1}{3!}A^3 + \frac{1}{5!}A^5 - \cdots \qquad (8.9)$$

is valid for all such matrices A.

Problem 8.1 Use Eq. (8.2) to prove: (a) if A is symmetric then so if $f(A)$; (b) if A is triangular then so is $f(A)$ (see Exercise 2.8); (c) if u_i is an eigenvector of A corresponding to λ_i, then $f(\lambda_i)$ and u_i are the corresponding eigenvalue and eigenvector of $f(A)$.

Problem 8.2 If

$$A = \begin{bmatrix} 0 & 2 & 1 \\ 0 & 0 & 3 \\ 0 & 0 & 0 \end{bmatrix} \qquad (8.10)$$

verify that $A^3 = 0$ and hence calculate $\cos A$ directly from (8.2).

Problem 8.3 Use Eq. (8.4) to calculate $\sin A$ for the matrix in (8.6).

It is important to realize that just because a matrix function is defined by a series like that for a scalar function, it does *not* follow that properties of the scalar function still apply. For example, the matrix A in (8.10) when substituted into (8.9) gives $\sin A = A$ (since $A^3 = A^5 = \cdots = 0$), but for scalar functions $\sin z = z$ only when $z = 0$.

To illustrate this point further, consider the important *exponential matrix* (or *matrix exponential*) $\exp A$ or e^A, defined by

$$e^A = I_n + A + \frac{1}{2!}A^2 + \frac{1}{3!}A^3 + \cdots \qquad (8.11)$$

which converges for all $n \times n$ matrices A having finite elements, since the series for e^λ converges for all finite λ. If B is a second $n \times n$ matrix then

$$e^{A+B} = I + (A + B) + \frac{1}{2!}(A + B)^2 + \cdots$$

$$= I + (A + B) + \frac{1}{2!}(A^2 + AB + BA + B^2) + \cdots \qquad (8.12)$$

and

$$(e^A)(e^B) = \left(I + A + \frac{1}{2!}A^2 + \cdots\right)\left(I + B + \frac{1}{2!}B^2 + \cdots\right)$$

$$= I + A + B + \frac{1}{2!}(A^2 + 2AB + B^2) + \cdots \qquad (8.13)$$

Comparison of (8.12) and (8.13) reveals that the relation

$$e^{A+B} = e^A e^B \qquad (8.14)$$

holds *only if* $AB = BA$. In general, formulae involving two or more matrices hold only if all the matrices commute with each other. For example,

$$\sin(A + B) = \sin A \cos B + \cos A \sin B \qquad (8.15)$$

only if A and B commute with each other. Formulae involving only a single matrix usually do carry over from the scalar case, for example,

$$\sin 2A = 2 \sin A \cos A$$

Problem 8.4 If

$$A = \begin{bmatrix} 1 & 1 \\ 0 & 0 \end{bmatrix}, \qquad B = \begin{bmatrix} 1 & -1 \\ 0 & 0 \end{bmatrix}$$

show that $A^2 = A$ and hence obtain $\exp A$ using (8.11). Similarly, obtain exp B. Determine also $\exp(A + B)$, $(\exp A)(\exp B)$, and $(\exp B)(\exp A)$ (notice that they are all different from each other).

Problem 8.5 Using (8.14) with $B = -A$, prove that $\exp A$ is nonsingular for any $n \times n$ matrix A. Hence deduce that

$$(\exp A)^{-1} = I - A + \tfrac{1}{2}A^2 - \cdots = \exp(-A).$$

Problem 8.6 Use (8.11) to prove that $(\exp A)^* = \exp(A^*)$. Hence show that if A is skew symmetric then $\exp A$ is orthogonal, and if A is skew hermitian then $\exp A$ is unitary.

Students are often disappointed that even simple results like (8.15) do not hold for matrix functions. However, as on previous occasions in this book, it is possible to use Kronecker products to overcome some of the snags arising from non-commutativity. First, if A is $n \times n$, then from (8.2)

$$
\begin{aligned}
f(A \otimes I_m) &= f_0 I_{nm} + f_1(A \otimes I_m) + f_2(A \otimes I_m)^2 + \cdots \\
&= f_0 I_n \otimes I_m + f_1(A \otimes I_m) + f_2(A^2 \otimes I_m) + \cdots \\
&= (f_0 I_n + f_1 A + f_2 A^2 + \cdots) \otimes I_m \\
&= f(A) \otimes I_m
\end{aligned}
\tag{8.16}
$$

(notice that we have used $(A \otimes I)^2 = A^2 \otimes I$, etc., easily proved using (2.75)). Similarly, it is left as a simple task for the reader to verify that if B is $m \times m$ then

$$f(I_n \otimes B) = I_n \otimes f(B) \tag{8.17}$$

Now consider the matrix D introduced in (5.53) and studied further in Section 6.3.5, namely

$$D = A \otimes I_m + I_n \otimes B \tag{8.18}$$

(the transpose on B has been dropped here for convenience, without affecting the argument). From the rule in (2.75) we have

$$(A \otimes I_m)(I_n \otimes B) \equiv (A I_n) \otimes (I_m B) \equiv (I_n \otimes B)(A \otimes I_m)$$

showing that $A \otimes I_m$ and $I_n \otimes B$ commute with each other. Thus scalar formulae do carry over for these matrices. For example,

$$
\begin{aligned}
\sin D &= \sin(A \otimes I_m + I_n \otimes B) \\
&= \sin(A \otimes I_m)\cos(I_n \otimes B) + \cos(A \otimes I_m)\sin(I_n \otimes B)
\end{aligned}
\tag{8.19}
$$

and using (8.16) and (8.17), the expression (8.19) can be simplified as follows:

$$\sin D = [(\sin A) \otimes I_m][I_n \otimes (\cos B)] + [(\cos A) \otimes I_m][I_n \otimes (\sin B)]$$
$$= (\sin A) \otimes (\cos B) + (\cos A) \otimes (\sin B) \tag{8.20}$$

again using (2.75). The formula (8.20) is seen to be a generalization of the usual scalar result for $\sin(a + b)$.

Problem 8.7 For the matrix D in (8.18) prove that

$$\exp D = (\exp A) \otimes (\exp B)$$

8.2 Sylvester's formula

A disadvantage of using Eq. (8.4) to evaluate $f(A)$ is the need to calculate the eigenvectors of A in order to obtain T. We now show how this can be avoided. As in (6.47), we can divide $f(\lambda)$ by the characteristic polynomial $k(\lambda)$ of A to give

$$f(\lambda) \equiv q(\lambda)k(\lambda) + \alpha(\lambda) \tag{8.21}$$

where the remainder polynomial $\alpha(\lambda)$ has degree at most $n - 1$. Since by definition $k(\lambda_r) = 0$ for each eigenvalue λ_r, substitution of $\lambda = \lambda_r$ into (8.21) gives

$$f(\lambda_r) = \alpha(\lambda_r)$$
$$= \alpha_0 + \alpha_1\lambda_r + \alpha_2\lambda_r^2 + \cdots + \alpha_{n-1}\lambda_r^{n-1} \tag{8.22}$$

for $r = 1, 2, \ldots, n$. Assuming all the eigenvalues of A are distinct, (8.22) represents n equations for the n unknown constants $\alpha_0, \ldots, \alpha_{n-1}$. In fact (8.22) can be regarded as determining the $(n - 1)$th degree polynomial $\alpha(\lambda)$ which takes the value $f(\lambda_r)$ when $\lambda = \lambda_r$, for each value of r (see Exercise 4.17). This is a standard problem of interpolation, and the desired (unique) polynomial is given by Lagrange's formula (Barnett and Cronin (1975), Formula 9.5.6):

$$\alpha(\lambda) = \sum_{r=1}^{n} f(\lambda_r)z_r , \quad z_r = \prod_{\substack{i=1 \\ i \neq r}}^{n} [(\lambda - \lambda_i)/(\lambda_r - \lambda_i)] \tag{8.23}$$

To obtain $f(A)$ we use the fact that by the Cayley–Hamilton theorem (6.37), $k(A) \equiv 0$, so on replacing λ by A in the identity (8.21) we obtain simply $f(A) \equiv \alpha(A)$. Hence, replacing λ by A in (8.23) gives *Sylvester's formula*

$$f(A) = \sum_{r=1}^{n} f(\lambda_r)Z_r \tag{8.24}$$

where the Z_r are now $n \times n$ matrices given by

$$Z_r = \prod_{\substack{i=1 \\ i \neq r}}^{n} [(A - \lambda_i I)/(\lambda_r - \lambda_i)] \tag{8.25}$$

Notice that (8.24) and (8.25) require a knowledge only of the eigenvalues of A, and that the matrices Z_r in (8.25) are determined entirely by A, i.e., the Z_r are independent of the function f.

Example 8.2 Let A be a 3×3 matrix having eigenvalues $\lambda_1 = 1$, $\lambda_2 = 2$, $\lambda_3 = -1$. Applying (8.25) with $n = 3$ we have

$$Z_1 = (A - \lambda_2 I)(A - \lambda_3 I)/(\lambda_1 - \lambda_2)(\lambda_1 - \lambda_3)$$

$$= -\frac{1}{2}(A - 2I)(A + I)$$

$$Z_2 = (A - \lambda_1 I)(A - \lambda_3 I)/(\lambda_2 - \lambda_1)(\lambda_2 - \lambda_3)$$

$$= \frac{1}{3}(A - I)(A + I)$$

$$Z_3 = (A - \lambda_1 I)(A - \lambda_2 I)/(\lambda_3 - \lambda_1)(\lambda_3 - \lambda_2)$$

$$= \frac{1}{6}(A - I)(A - 2I)$$

Then for *any* function $f(\lambda)$ for which the convergence condition is satisfied we have in this case

$$f(A) = f(1)Z_1 + f(2)Z_2 + f(-1)Z_3$$

with the matrices Z_r given above.

Example 8.3 Return to the matrix A in (8.6). We have

$$Z_1 = (A - \lambda_2 I)/(\lambda_1 - \lambda_2) = \left(A - \frac{1}{2}I\right) \Big/ \left(\frac{1}{4} - \frac{1}{2}\right)$$

$$= \begin{bmatrix} 2 & -4 \\ \frac{1}{2} & -1 \end{bmatrix}$$

$$Z_2 = (A - \lambda_1 I)/(\lambda_2 - \lambda_1) = \left(A - \frac{1}{4}I\right) \Big/ \left(\frac{1}{2} - \frac{1}{4}\right)$$

$$= \begin{bmatrix} -1 & 4 \\ -\frac{1}{2} & 2 \end{bmatrix}$$

Again, if $f(A)$ exists, then $f(A) = f(\frac{1}{4})Z_1 + f(\frac{1}{2})Z_2$. For example, if $f(\lambda) = \ln(1 + \lambda)$, then

$$\ln(I_2 + A) = (\ln \tfrac{5}{4})Z_1 + (\ln \tfrac{3}{2})Z_2 \tag{8.26}$$

and on substituting for Z_1 and Z_2 it can be confirmed that (8.26) agrees with (8.8).

Notice that it is easily proved (see Exercise 8.5) that the matrices Z_r in (8.25) satisfy

$$\sum_{r=1}^{n} Z_r = I_n, \qquad Z_r Z_s = 0 \ (r \neq s) \tag{8.27}$$

and these provide a simple numerical check on the calculations. The reader can easily confirm that in the preceding example $Z_1 + Z_2 = I_2$, $Z_1 Z_2 = 0$.

Problem 8.8 Verify that (8.27) holds for the matrices Z_1, Z_2, Z_3 in Example 8.2.

Example 8.9 Repeat Problem 8.3, using Z_1 and Z_2 in Example 8.3.

Problem 8.10 If

$$A = \begin{bmatrix} 3 & 2 \\ 2 & 3 \end{bmatrix}$$

use Sylvester's formula to determine $\sin A$. Similarly, write down an expression for A^{100}.

Problem 8.11 For the matrix A in Problem 6.8b calculate $\exp A$: (a) using Sylvester's formula; (b) using Eq. (8.4) and the similarity transformation obtained in the solution of Problem 6.28.

8.3 Linear differential and difference equations

We return to the systems of equations studied in Section 6.5, and show how they can be solved using matrix functions. First consider the differential equations (6.91), i.e.,

$$\dot{x} = Ax \tag{8.28}$$

where A is a constant $n \times n$ matrix. If a is a scalar, then the solution of $\dot{x} = ax$ is $x(t) = \exp(at)x(0)$. This suggests that we consider

$$\exp(At) = I_n + tA + \frac{t^2}{2!}A^2 + \frac{t^3}{3!}A^3 + \cdots \tag{8.29}$$

where t represents time. The series (8.29) converges for all finite values of t. Differentiating (8.29):

$$\frac{d}{dt}(\exp(At)) = 0 + A + \frac{2t}{2!}A^2 + \frac{3t^2}{3!}A^3 + \cdots$$

$$= A\left(I + tA + \frac{t^2}{2!}A^2 + \cdots\right) = A\exp(At)$$

which is the same result as for scalar exponentials. It follows by analogy with the scalar case that the solution of (8.28) can be written

$$x(t) = \exp(At)x_0 \tag{8.30}$$

where x_0 is the value of $x(t)$ when $t = 0$ (as can be verified by checking that (8.30) does indeed satisfy (8.28)).

Problem 8.12 Prove that $\exp(At_1)\exp(At_2) = \exp[A(t_1 + t_2)]$ for any finite values of t_1 and t_2.

Problem 8.13 Determine $\exp(At)$ for the matrix A in (8.10).

An advantage of using (8.30) instead of (6.96) is that there is no need to calculate eigenvectors of A, since $\exp(At)$ can be obtained using Sylvester's formula.

Example 8.4 Return to the problem in Example 6.13 where A is the matrix in (6.5) and $\lambda_1 = -1$, $\lambda_2 = 4$. From (8.25) with $n = 2$,

$$Z_1 = -\frac{1}{5}\begin{bmatrix} -3 & 3 \\ 2 & -2 \end{bmatrix}, \qquad Z_2 = \frac{1}{5}\begin{bmatrix} 2 & 3 \\ 2 & 3 \end{bmatrix} \tag{8.31}$$

and from (8.24) with $f(\lambda) = e^{\lambda t}$, we have $e^{At} = e^{-t}Z_1 + e^{4t}Z_2$. Then (8.30) gives, with $x_0 = [x_{10}, x_{20}]^T$, the solution of (8.28) in this case:

$$x(t) = \frac{1}{5}\begin{bmatrix} (2e^{4t} + 3e^{-t}) & 3(e^{4t} - e^{-t}) \\ 2(e^{4t} - e^{-t}) & (3e^{4t} + 2e^{-t}) \end{bmatrix}\begin{bmatrix} x_{10} \\ x_{20} \end{bmatrix}$$

It is readily confirmed that this agrees with the solution previously obtained in (6.98), with appropriate expressions for the constants α_1 and α_2 in terms of x_{10} and x_{20}.

Problem 8.14 Solve (8.28) using (8.30) when A is the matrix in Problem 6.8b (compare your answer with that for Problem 6.36).

It is interesting to consider Laplace transformation of (8.28). As in Exercise 4.8, this gives

$$s\bar{x} - x_0 = A\bar{x}$$

where $\bar{x} = \mathcal{L}\{x(t)\}$, and on rearrangement this becomes

$$\bar{x} = (sI_n - A)^{-1}x_0 \tag{8.32}$$

However, from (8.30)

$$\bar{x} = \mathcal{L}\{e^{At}x_0\} \tag{8.33}$$

and since x_0 is an arbitrary constant vector, it follows from (8.32) and (8.33) that

$$\mathscr{L}\{e^{At}\} = (sI - A)^{-1}$$

which is a generalization of the well-known result for scalars: $\mathscr{L}\{e^{at}\} = 1/(s - a)$ (Barnett and Cronin (1975), Formula 6.4.7).

Turning to the linear difference equations in (6.99), i.e.

$$X(k + 1) = AX(k), \qquad k = 0, 1, 2, \ldots \qquad (8.34)$$

the solution can in this case be written

$$X(k) = A^k X(0) \qquad (8.35)$$

This is readily verified, since (8.35) gives $X(k + 1) = A^{k+1}X(0) = AA^k X(0) = AX(k)$ as required. Evaluation of A^k can again be carried out by Sylvester's formula (8.24) by setting $f(\lambda) = \lambda^k$, thus avoiding the calculation of eigenvectors of A needed in (6.101).

Problem 8.15 Solve the difference equations in Example 6.14 using (8.35) and Sylvester's formula (Z_1 and Z_2 are given in (8.31)).

Exercises

8.1 If A is a block diagonal matrix, i.e.,

$$A = \text{diag}[A_1, A_2, \ldots, A_k]$$

where each A_i is a square matrix (e.g., see (2.64)) prove that $f(A)$ is also block diagonal, and

$$f(A) = \text{diag}[f(A_1), f(A_2), \ldots, f(A_k)].$$

8.2 Let A be a 4×4 matrix having eigenvalues $\pm k$, 0, 0. Use the Cayley–Hamilton theorem to show that $A^4 = k^2 A^2$. Hence, using the series definition for $\cos A$, prove that $\cos A = I + (\cos k - 1)A^2/k^2$.

8.3 Prove using Eq. (8.4) that

$$\det f(A) = f(\lambda_1)f(\lambda_2) \cdots f(\lambda_n).$$

Hence show that $\det(\exp A) = \exp(\text{tr} A)$, and deduce that when A is real and skew symmetric then $\det(\exp A) = 1$.

8.4 Write the equation of simple harmonic motion

$$\ddot{z} + \omega^2 z = 0 \qquad (8.36)$$

in the form (8.28) by taking $x_1 = z$, $x_2 = \dot{z}/\omega$. By calculating A^2 and using (8.29), show that

$$e^{At} = \begin{bmatrix} \cos \omega t & \sin \omega t \\ -\sin \omega t & \cos \omega t \end{bmatrix}$$

and hence determine the solution $z(t)$ of (8.36).

8.5 By suitable choices of $f(\lambda)$ in (8.24) prove:

$$\text{(a)} \sum_{r=1}^{n} Z_r = I_n; \qquad \text{(b)} \sum_{r=1}^{n} \lambda_r^k Z_r = A^k, \qquad k = 1, 2, 3, \ldots$$

8.6 For the matrix A in (2.25) use the result of Problem 2.9 to prove by induction that

$$A^k = (2^k - 1)A - (2^k - 2)I_3$$

for any positive integer k. Hence show that

$$\exp A = (e^2 - e)A - (e^2 - 2e)I_3.$$

8.7 If A is a real $n \times n$ matrix prove that

$$\exp(iA) = \cos A + i \sin A$$

and hence show that

$$\cos A = \frac{1}{2}[\exp(iA) + \exp(-iA)], \qquad \sin A = \frac{1}{2i}[\exp(iA) - \exp(-iA)].$$

8.8 Consider again the control system Eqs (4.36). Use the transformation $z(t) = \exp(-At)x(t)$ and the product rule (2.70) to obtain a differential equation for $z(t)$. Hence show that the general solution of (4.36) subject to $x(0) = x_0$ can be written

$$x(t) = e^{At}\left[x_0 + \int_0^t e^{-A\tau} bu \, d\tau\right].$$

Bibliography

References in text

Barnett, S., and T. M. Cronin (1975) *Mathematical formulae for engineering and science students* (2nd ed.), Bradford University Press – Crosby Lockwood Staples, London.

Bellman, R. (1970) *Introduction to matrix analysis* (2nd ed.), McGraw-Hill, New York.

Fox, L. (1964) *An introduction to numerical linear algebra*, Clarendon Press, Oxford.

Goult, R. J., R. F. Hoskins, J. A. Milner, and M. J. Pratt (1974) *Computational methods in linear algebra*, Stanley Thornes, London.

Jones, L. B. (1976) *Ordinary differential equations for engineering and science students*, Bradford University Press – Crosby Lockwood Staples, London.

Mirsky, L. (1963) *An introduction to linear algebra*, Clarendon Press, Oxford.

Stewart, G. W. (1973) *Introduction to matrix computations*, Academic Press, New York.

Some books on applications

Almon, C. (1967) *Matrix methods in economics*, Addison-Wesley, Reading, Mass.

Atkins, P. W. (1970) *Molecular quantum mechanics*, Clarendon Press, Oxford.

Barnett, S. (1971) *Matrices in control theory, with applications to linear programming*, Van Nostrand Reinhold, London.

Barnett, S. (1975) *Introduction to mathematical control theory*, Clarendon Press, Oxford.

Barnett, S., and C. Storey (1970) *Matrix methods in stability theory*, Nelson, London.

Braae, R. (1963) *Matrix algebra for electrical engineers*, Pitman, London.

Brand, T., and A. Sherlock (1970) *Matrices: pure and applied*, Arnold, London.

Brouwer, W. (1964) *Matrix methods in optical instrument design*, Benjamin, New York.

175

Chen, C. T. (1970) *Introduction to linear system theory*, Holt, Rinehart and Winston, New York.

Eisenschitz, R. K. (1966) *Matrix algebra for physicists*, Heinemann, London.

Fletcher, T. J. (1972) *Linear algebra through its applications*, Van Nostrand Reinhold, London.

Gass, S. I. (1975) *Linear programming* (4th ed.), McGraw-Hill, New York.

Graybill, F. A. (1969) *Introduction to matrices with applications in statistics*, Wadsworth, Belmont, California.

Hancock, N. N. (1974) *Matrix analysis of electrical machinery* (2nd ed.), Pergamon, Oxford.

Horst, P. (1963) *Matrix algebra for social scientists*, Holt, Rinehart and Winston, New York.

Jury, E. I. (1974) *Inners and stability of dynamic systems*, Wiley-Interscience, New York.

Kemeny, J. G., J. L. Snell, and G. L. Thompson (1974) *Introduction to finite mathematics* (3rd ed.), Prentice-Hall, New Jersey.

Noble, B., and J. Daniel (1977) *Applied linear algebra* (2nd ed.), Prentice-Hall, New Jersey.

Pestel, E. C., and F. A. Leckie (1963) *Matrix methods in elastomechanics*, McGraw-Hill, New York.

Pipes, L. A. (1963) *Matrix methods for engineering*, Prentice-Hall, New Jersey.

Pipes, L. A., and S. A. Hovanessian (1969) *Matrix computer methods for engineering*, Wiley, New York.

Robinson, J. (1966) *Structural matrix analysis for the engineer*, Wiley, New York.

Rorres, C. and H. Anton (1977) *Applications of linear algebra*, Wiley, New York.

Rubinstein, M. F. (1966) *Matrix computer analysis of structures*, Prentice-Hall, New Jersey.

Searle, S. R. (1966) *Matrix algebra for the biological sciences*, Wiley-Interscience, New York.

Seshu, S., and M. B. Reed (1961) *Linear graphs and electrical networks*, Addison-Wesley, Reading, Mass.

Thomson, W. T. (1972) *Theory of vibrations with applications*, Prentice-Hall, New Jersey.

Answers to problems and exercises

Chapter 1

Exercises

1.1 $\quad \begin{array}{c} \\ e_1 \\ e_2 \\ e_3 \end{array} \begin{array}{c} f_1 \; f_2 \\ \begin{bmatrix} 1 & 0 \\ 1 & 1 \\ 0 & 1 \end{bmatrix} \end{array}$

1.2 (a) $\begin{array}{c} \\ B_1 \\ B_2 \\ B_3 \end{array} \begin{array}{c} C_1 \; C_2 \; C_3 \\ \begin{bmatrix} 0 & 1 & 2 \\ 2 & 3 & 0 \\ 0 & 0 & 1 \end{bmatrix} \end{array}$ (b) $\begin{array}{c} \\ A_1 \\ A_2 \end{array} \begin{array}{c} C_1 \; C_2 \; C_3 \\ \begin{bmatrix} 2 & 7 & 11 \\ 4 & 6 & 2 \end{bmatrix} \end{array}$

1.3 $\quad A \begin{array}{c} H \\ T \end{array} \begin{array}{c} H \quad\; T \\ \begin{bmatrix} 5 & -2 \\ 3 & -7 \end{bmatrix} \end{array}, + = $ money from A to B

1.4 $\quad \begin{bmatrix} 2 & 8 & 0 & 2 & -1 & 0 & 0 \\ 1 & 2 & 28 & 9 & 0 & -1 & 0 \\ 16 & 0 & 0 & 1 & 0 & 0 & -1 \end{bmatrix} \begin{bmatrix} x_1 \\ x_2 \\ \vdots \\ x_7 \end{bmatrix} = \begin{bmatrix} 2000 \\ 2500 \\ 8500 \end{bmatrix}$, each $x_i \geqslant 0$

1.5 (a) reflection in x_1-axis (b) stretch (c) rotation about x_3-axis

Chapter 2

Problems

2.1 (a) 4×2 (b) 3, 5, undefined

2.2 (a) $\begin{bmatrix} 1 & 0 & -1 \\ 3 & 2 & 1 \end{bmatrix}$ (b) $\begin{bmatrix} 0 & 1 & 2 \\ 1 & 0 & 1 \\ 2 & 1 & 0 \end{bmatrix}$

2.3 (a) $\begin{bmatrix} 2 & 2 \\ 7 & 3 \\ 9 & 6 \\ 5 & -14 \end{bmatrix}$ (b) $\begin{bmatrix} 0 & -4 \\ -7 & 5 \\ -1 & -2 \\ 1 & -2 \end{bmatrix}$ (c) $\begin{bmatrix} 2 & 6 \\ 14 & -2 \\ 10 & 8 \\ 4 & -12 \end{bmatrix}$ (d) $\begin{bmatrix} -2 & 10 \\ 14 & -18 \\ -6 & 0 \\ -8 & 20 \end{bmatrix}$

2.5 49 units vit. A, 109 units vit. B, 1205p.

177

2.6 (a) $A + B = \begin{bmatrix} 0 & 0 \\ 2 & 1 \end{bmatrix}$, $\quad AB = \begin{bmatrix} 1 & 0 \\ 1 & -1 \end{bmatrix}$, $\quad BA = \begin{bmatrix} -1 & -1 \\ 0 & 1 \end{bmatrix}$

(b) $A + B = \begin{bmatrix} 2 & 3 & 4 \\ 2 & 5 & 9 \end{bmatrix}$

(c) $AB = \begin{bmatrix} -1 & -8 & -10 \\ 1 & -2 & -5 \\ 9 & 22 & 15 \end{bmatrix}$, $\quad BA = \begin{bmatrix} 15 & -21 \\ 10 & -3 \end{bmatrix}$

2.7 $m = 3$, $n = 8$ $\quad$ **2.10** $a_2 = b_1$, $b_2 = c_1, \ldots, y_2 = z_1$; $a_1 \times z_2$

2.19 ith row, column of A

2.20 $M = \begin{bmatrix} 3 & 0 & 0 \\ 0 & 1 & -1 \\ 0 & -1 & 1 \end{bmatrix}$, $\quad S = \begin{bmatrix} 0 & -1 & -1 \\ 1 & 0 & 0 \\ 1 & 0 & 0 \end{bmatrix}$

2.23 $\frac{1}{2}n(n-1)$ $\quad$ **2.25** (b) $A_1A_2 = -A_2A_1$

2.28 (R1), (C1); (R2), (C2)

2.31 $A^T = \mathrm{diag}[A_1^T, A_2^T]$, $A^2 = \mathrm{diag}[A_1^2, A_2^2]$

2.35 $\dot{A}A + A\dot{A}$, $\dot{A}A^2 + A\dot{A}A + A^2\dot{A}$

Exercises

2.2 $v_1 = (-31 + 16i)v_3 + (-28 + 24i)i_3$, $i_1 = (-32 + 8i)v_3 + (-31 + 16i)i_3$

2.5 $\begin{bmatrix} 2 & 3 & 0 \\ 0 & 0 & 4 \\ 1 & -7 & 3 \end{bmatrix}$ $\quad$ **2.11** 21, 34

Chapter 3

Problems

3.1 $x_1 = (a_{22}b_1 - a_{12}b_2)/d$, $x_2 = (a_{11}b_2 - a_{21}b_1)/d$, $d = a_{11}a_{22} - a_{12}a_{21}$
(a) $d \neq 0$ (b) $d = 0$, $a_{11}/a_{21} = b_1/b_2$ (c) $d = 0$, $a_{11}/a_{21} \neq b_1/b_2$

3.2 $(1, 13/5, 11/5)$ $\quad$ **3.3** $(3, 4, 5)$

3.6 $(1, -2, -1, 1)$

3.7 $\begin{bmatrix} 1 & 0 & 0 \\ 2 & 1 & 0 \\ 3 & 2 & 1 \end{bmatrix}\begin{bmatrix} 2 & 3 & 4 \\ 0 & 4 & 1 \\ 0 & 0 & 6 \end{bmatrix}$, $(5, 3, 1)$

3.9 $(1, 2, 3)$, $(3, -1, 2)$

3.11 $U = \frac{1}{2}\begin{bmatrix} 4 & 1 & 3 \\ 0 & 3 & -1 \\ 0 & 0 & 5\sqrt{2} \end{bmatrix}$

3.12 (a) $(0.609, -0.739)$ (b) $(-28.913, -43.891)$

ANSWERS TO PROBLEMS AND EXERCISES

Exercises

3.1 $(1.26, 0.05, -0.33)$

3.2 $a_1 \neq 0$, $d_1 = a_1a_2 - b_1c_1 \neq 0$, $d_2 = a_3d_1 - a_1b_2c_2 \neq 0$, $a_4d_2 - b_3c_3d_1 \neq 0$

3.3 $l_1 = 2$, $l_2 = -3/5$, $l_3 = -35/12$, $u_1 = 1$, $u_2 = -5$, $u_3 = -12/5$,
$u_4 = 13/12$, $v_1 = 2$, $v_2 = 1$, $v_3 = -1$, $x = (2, 3, -2, 5)$

3.4 $x_1^2 + x_2^2 - 6x_1 - 4x_2 - 12 = 0$

3.5 $\begin{bmatrix} 2 & 5/2 \\ -4 & -9/2 \\ -1 & -2 \end{bmatrix}$

Chapter 4

Problems

4.2 -55 **4.5** -45

4.12 $2\mathbf{i} - 3\mathbf{j} - 5\mathbf{k}, \frac{1}{2}\sqrt{38}$

4.14 $R_1R_2C \neq L$ **4.15** $25, -27$

4.16 (a) 7   (b) 63

4.19 $a_{1n}a_{2,n-1} \ldots a_{n1}d$; $\det J_n = d = (-1)^{n/2}$, n even; $d = (-1)^{(n-1)/2}$, n odd

4.21 $\dfrac{1}{17}\begin{bmatrix} 6 & -3 & 14 \\ -5 & 11 & -23 \\ -2 & 1 & 1 \end{bmatrix}$ **4.22** $(2, -3, 0)$

4.25 A and B commute   **4.29** $\dfrac{1}{289}\begin{bmatrix} 65 & -75 & 197 \\ -75 & 131 & -294 \\ 197 & -294 & 726 \end{bmatrix}$

4.31 $\dfrac{1}{25}\begin{bmatrix} 55 & -25 & 5 \\ 24 & -10 & -1 \\ -2 & 5 & -2 \end{bmatrix}$ **4.32** $\dfrac{1}{14}\begin{bmatrix} 0 & 4 & 2 & -6 \\ 21 & -9 & -8 & 3 \\ 7 & -1 & -4 & 5 \\ -14 & 6 & 10 & -2 \end{bmatrix}$

Exercises

4.2 $k < 2$

4.6 $\begin{bmatrix} A^{-1} & 0 \\ -D^{-1}CA^{-1} & D^{-1} \end{bmatrix}$ **4.13** $A^5 = \begin{bmatrix} 122 & 121 \\ 121 & 122 \end{bmatrix}$

4.26 204

Chapter 5

Problems

5.2 (a) $k = 8$ (b) $k \neq 8$ (c) none

5.4 2   **5.5** (a) 2   (b) 2   (c) 3

5.10 (a) yes in all cases

(b) and (c) yes if nonsingular, no if singular

5.11 (a) $x_1 = \frac{7}{2}t_1 - \frac{5}{2}t_2$, $x_2 = -3t_1 + 2t_2$, $x_3 = t_1$, $x_4 = t_2$

(b) $x_1 = 2t_1$, $x_2 = -4t_1$, $x_3 = t_1$

(c) $x_1 = t_1 - t_3$, $x_2 = -4t_1 - 4t_2 - 2t_3$, $x_3 = t_1$, $x_4 = 0$, $x_5 = t_2$, $x_6 = t_3$

5.12 (a) $x_1 = -2t_1$, $x_2 = t_1$, $x_3 = 0$ (b) no

5.13 consistent **5.14** $k = 6$

5.15 $x_1 = 22 - 7t_1 + 5t_2$, $x_2 = 6 - t_1 + t_2$, $x_3 = t_1$, $x_4 = t_2$

5.16 $x_1 = -9 - 2t_1$, $x_2 = 15 + t_1$, $x_3 = t_1$

5.17 (a) $k = 3$ (b) $k = 1$ **5.18** $y = \frac{1}{2} + \frac{8}{5}x$

5.20 $X = \frac{1}{238}\begin{bmatrix} 95 & -33 \\ -7 & 35 \end{bmatrix}$

5.22 (a) $7(C1) - 6(C2) + 2(C3) \equiv 0$, $(R1) + 2(R2) - (R3) \equiv 0$

(b) $2(C1) - 4(C2) + (C3) \equiv 0$, $5(R1) - (R3) - 2(R4) \equiv 0$

5.23 $b = \frac{1}{3}(4b_1 + b_2 - 2b_3)g_1 + \frac{1}{3}(-b_1 - b_2 + 2b_3)g_2 + \frac{1}{3}(-2b_1 + b_2 + b_3)g_3$

Exercises

5.1 not controllable

5.2 $x_{11} = -2 + t_1 + t_2$, $x_{12} = 12 - t_1$, $x_{13} = 7 - t_2$, $x_{21} = 11 - t_1 - t_2$, $x_{22} = t_1$, $x_{23} = t_2$

5.3 $x_1 = 2 - t_1 - t_2$, $x_2 = -t_1 + 2t_2$, $x_3 = t_1$, $x_4 = t_2$

5.4 $k = 2$, $x_1 = -t_1$, $x_2 = t_1$, $x_3 = t_1$

$k = 5$, $x_1 = \frac{1}{2}t_2$, $x_2 = \frac{1}{4}t_2$, $x_3 = t_2$

5.5 $\lambda = -1$, $x_1 = -1/11$, $x_2 = -15/11$

$\lambda = 1$, $x_1 = -5$, $x_2 = 1$; $\lambda = 12$, $x_1 = \frac{1}{2}$, $x_2 = 1$

5.6 $a = 0$, $b = \frac{7}{2}$, $R(A) = 2$, $x_1 = -3t_1 - \frac{5}{2}t_2$, $x_2 = 2t_1 + \frac{3}{2}t_2$, $x_3 = t_1$, $x_4 = t_2$;

$R(A) = 3$ if $a \neq 0$ (any b) or $b \neq \frac{7}{2}$ (any a), and $x_1 = -3t_1$, $x_2 = 2t_1$, $x_3 = t_1$, $x_4 = 0$

Chapter 6

Problems

6.1 $4, (1/\sqrt{2})[1, 1]^{T}; -2, (1/\sqrt{2})[1, -1]^{T}$

6.6 (a) $-2 \pm i\sqrt{(21)}$ (b) $4, \pm 2i$

6.7 (a) $\frac{1}{3}[1, 2, 2]^{T}, \frac{1}{3}[2, 1, -2]^{T}$

(b) $(1/\sqrt{59})[-1, 7, 3]^{T}, (1/\sqrt{3})[i, 1, -1]^{T}, (1/\sqrt{3})[i, -1, 1]^{T}$

6.8　(a) $0, 3 \pm 3\sqrt{3}$　(b) $1, 2, 3$

6.9　$1, (1/\sqrt{6})[-(1+i), 2]^T; 4, (1/\sqrt{3})[(1+i), 1]^T$

6.18　$\lambda^3 - 5\lambda^2 + 8\lambda - 4$

6.19　$A^4 = \begin{bmatrix} -49 & -50 & -40 \\ 65 & 66 & 40 \\ 130 & 130 & 81 \end{bmatrix}$, $A^{-1} = \frac{1}{6}\begin{bmatrix} 4 & -2 & 2 \\ -1 & 5 & -2 \\ -2 & -2 & 2 \end{bmatrix}$

6.20　$\begin{bmatrix} -15 & -8 & 0 \\ 8 & 17 & 24 \\ 32 & 40 & 49 \end{bmatrix}$　**6.22**　$\begin{bmatrix} -2 & -1 & 1 \\ -6 & -13 & -7 \\ 42 & 71 & 29 \end{bmatrix}$, yes

6.23　$A^2 \otimes I_m + I_n \otimes B^2$　**6.24**　$\lambda_i \mu_j \nu_k$, $u_i \otimes y_j \otimes z_k$

6.27　similar in each case　**6.28**　$[1, -1, 0]^T, [2, -1, -2]^T, [1, -1, -2]^T$

6.29　$A^3 = \begin{bmatrix} -11 & -12 & -13 \\ 19 & 20 & 13 \\ 38 & 38 & 27 \end{bmatrix}$　**6.32**　9, 18

6.33　n

6.36　$x_1 = \alpha_1 e^t + 2\alpha_2 e^{2t} + \alpha_3 e^{3t}$, $x_2 = -\alpha_1 e^t - \alpha_2 e^{2t} - \alpha_3 e^{3t}$,
$x_3 = -2\alpha_2 e^{2t} - 2\alpha_3 e^{3t}$

6.39　$\lambda_1 = 6, u_1 = [\frac{1}{2}, 1]^T, \lambda_2 = 3$

6.42　(a) $x^{(3)} = [2.120, 1.414, -1.236]^T$
(b) $x^{(3)} = [2.055, 1.469, -1.318]^T$
Exact solution is $x = [2, 3/2, -4/3]^T$

6.43　$-2 < a < 2, -4 < b < 2$

6.44　$X_1 = \begin{bmatrix} 0.09 & 0.18 \\ 0.27 & -0.46 \end{bmatrix}$, $X_2 = \begin{bmatrix} 0.091 & 0.182 \\ 0.273 & -0.455 \end{bmatrix}$

Exercises

6.6　$\lambda^4 - \lambda^3 - 13\lambda^2 + 20\lambda + 13$　　**6.7**　$\frac{1}{2}(3 \pm \sqrt{5}), \frac{1}{2}(5 \pm \sqrt{5})$

6.13　$k_n \lambda^n + k_{n-1}\lambda^{n-1} + \cdots + k_1\lambda + 1 = 0$　**6.18**　$\frac{1}{9}\begin{bmatrix} 14 & -2 & -14 \\ -2 & -1 & -16 \\ -14 & -16 & 5 \end{bmatrix}$

6.26　$\lambda_1 = 16, u_1 = [\frac{1}{2}, 1, -\frac{1}{2}]^T$　**6.27**　4, 2

6.29　-3.56　　**6.30**　(0.52, 0.27, 0.48, 0.73, 0.55, 0.12, 0.69, 0.35, 0.84)

Chapter 7

Problems

7.2　$2x_1^2 + 4x_1x_2 - x_2^2 + 5x_1x_3 - x_2x_3 + 4x_3^2$

7.3　$2|x_1|^2 + (1+i)\bar{x}_1 x_2 + (1-i)x_1\bar{x}_2 + 7|x_2|^2 + (5-i)\bar{x}_1 x_3$
$+ (5+i)x_1\bar{x}_3 + i\bar{x}_2 x_3 - ix_2\bar{x}_3 - |x_3|^2$

7.7 3

7.8 $2y_1^2 + y_2^2 + \frac{5}{8}y_3^2$; $x_1 = y_1 + y_2 + \frac{15}{4}y_3$, $x_2 = y_2 + \frac{5}{2}y_3$, $x_3 = y_3$

7.9 (a) $2\left(x_3 + \frac{5}{4}x_2\right)^2 - \frac{25}{8}\left(x_2 - \frac{12}{25}x_1\right)^2 + \frac{18}{25}x_1^2$

(b) $\left(z_1 + \frac{1}{2}z_2\right)^2 - \frac{1}{4}(z_2 - 2z_3)^2 + z_3^2$, $x_1 = z_1$, $x_2 = z_1 + z_2$, $x_3 = z_3$

7.10 3, 1   **7.11** positive semidefinite

7.14 positive definite, indefinite, indefinite

7.20 (a) negative semidefinite   (b) indefinite

7.21 (a) indefinite   (b) positive semidefinite   (c) positive definite

7.29 $P = \frac{1}{30}\begin{bmatrix} 13 & -1 \\ -1 & 4 \end{bmatrix}$, yes

Exercises

7.2 positive semidefinite if $R(A) < n$, positive definite if $R(A) = n$

7.4 (a) $3 > k > 2$ (b) $k > 25$ **7.7** $[1, -2, -2]^{\mathrm{T}}$

7.12 positive definite

Chapter 8

Problems

8.2 $\begin{bmatrix} 1 & 0 & -3 \\ 0 & 1 & 0 \\ 0 & 0 & 1 \end{bmatrix}$

8.3 $\begin{bmatrix} \left(2\sin\frac{1}{4} - \sin\frac{1}{2}\right) & 4\left(\sin\frac{1}{2} - \sin\frac{1}{4}\right) \\ \frac{1}{2}\left(\sin\frac{1}{4} - \sin\frac{1}{2}\right) & \left(2\sin\frac{1}{2} - \sin\frac{1}{4}\right) \end{bmatrix}$

8.4 $e^A = \begin{bmatrix} e & (e-1) \\ 0 & 1 \end{bmatrix}$, $e^B = \begin{bmatrix} e & (1-e) \\ 0 & 1 \end{bmatrix}$, $e^{A+B} = \begin{bmatrix} e^2 & 0 \\ 0 & 1 \end{bmatrix}$

8.10 $\sin A = \frac{1}{2}\begin{bmatrix} (\sin 5 + \sin 1) & (\sin 5 - \sin 1) \\ (\sin 5 - \sin 1) & (\sin 5 + \sin 1) \end{bmatrix}$, $A^{100} = \frac{1}{2}\begin{bmatrix} (5^{100} + 1) & (5^{100} - 1) \\ (5^{100} - 1) & (5^{100} + 1) \end{bmatrix}$

8.11 $\begin{bmatrix} (2e^2 - e^3) & (2e^2 - e - e^3) & \frac{1}{2}(e - e^3) \\ (e^3 - e^2) & (e - e^2 + e^3) & \frac{1}{2}(e^3 - e) \\ (2e^3 - 2e^2) & (2e^3 - 2e^2) & e^3 \end{bmatrix}$

8.13 $\begin{bmatrix} 1 & 2t & (t + 3t^2) \\ 0 & 1 & 3t \\ 0 & 0 & 1 \end{bmatrix}$

Exercises

8.4 $z(0)\cos\omega t + (\dot{z}(0)\sin\omega t)/\omega$

Index

Note entries under 'matrix'.

INDEX

Deficiency problem:

$$A = \begin{bmatrix} 5 & -3 & -9 \\ -2 & 2 & 4 \\ 2 & -1 & -3 \end{bmatrix} \qquad \lambda = 2 \qquad \begin{bmatrix} 2 \\ -1 \\ 1 \end{bmatrix}, 1 \qquad \begin{bmatrix} 3 \\ -2 \\ 2 \end{bmatrix}$$

$$\boxed{\begin{bmatrix} 2 \\ -1 \\ 1 \end{bmatrix} e^{2t}, \begin{bmatrix} 3 \\ -2 \\ 2 \end{bmatrix} e^{t}} \qquad \text{deficiency}$$

$$\lambda = 1 \qquad Z_2 = \begin{bmatrix} 3 \\ -2 \\ 2 \end{bmatrix}$$

$$(A - \lambda I) = \begin{bmatrix} 4 & -3 & -9 & | & 3 \\ -2 & 1 & 4 & | & -2 \\ 2 & -1 & -4 & | & 2 \end{bmatrix} \longrightarrow \begin{bmatrix} 2 & 1 & -4 & | & 2 \\ 0 & 1 & 1 & | & 1 \\ 0 & 0 & 0 & | & 0 \end{bmatrix}$$

choose $x_3 = 0$

$$Z_1 = \begin{bmatrix} 3/2 \\ 1 \\ 0 \end{bmatrix} \qquad \boxed{X_3(t) = \left(\begin{bmatrix} 3/2 \\ 1 \\ 0 \end{bmatrix} + \begin{bmatrix} 3 \\ -2 \\ 2 \end{bmatrix} t \right) e^{t}}$$

The 3 fund. sets.

$$x = (Z_1 + Z_2 t) e^{\lambda t}$$

VARIATION of parms:

$$dx/dt = \begin{bmatrix} 0 & 1 \\ -1 & 0 \end{bmatrix} x + \begin{bmatrix} 0 \\ \sin t \end{bmatrix}$$

$$X(t) = \begin{bmatrix} \cos t & \sin t \\ -\sin t & \cos t \end{bmatrix}$$

$$X^{-1}(t) g(t) = \begin{bmatrix} \cos & -\sin \\ \sin & \cos \end{bmatrix} \begin{bmatrix} 0 \\ \sin \end{bmatrix} = \begin{bmatrix} -\sin^2 t \\ \sin t \cos t \end{bmatrix}$$

$$v(t) = \int X^{-1} g(t) dt = \int \begin{bmatrix} -\sin^2 t \\ \sin \cos \end{bmatrix} \quad x_p = \frac{1}{2} \begin{bmatrix} \sin t - t\cos t \\ t \sin t \end{bmatrix} = X(t) v(t)$$

$$X(t) = \begin{bmatrix} \sin & -t\cos \\ t\sin t \end{bmatrix} + \begin{bmatrix} \cos & -\sin \\ \sin & \cos \end{bmatrix} \begin{bmatrix} c_1 \\ c_2 \end{bmatrix}$$